Synthesis Lectures on Human Language Technologies

Series Editor

Graeme Hirst, Department of Computer Science, University of Toronto, Toronto, ON, Canada

The series publishes topics relating to natural language processing, computational linguistics, information retrieval, and spoken language understanding. Emphasis is on important new techniques, on new applications, and on topics that combine two or more HLT subfields.

Lisa Beinborn · Nora Hollenstein

Cognitive Plausibility in Natural Language Processing

 Springer

Lisa Beinborn
Faculty of Humanities
Vrije Universiteit Amsterdam
Amsterdam, The Netherlands

Nora Hollenstein
Department of Nordic Studies and Linguistics
University of Copenhagen
Copenhagen, Denmark

ISSN 1947-4040   ISSN 1947-4059 (electronic)
Synthesis Lectures on Human Language Technologies
ISBN 978-3-031-43259-0   ISBN 978-3-031-43260-6 (eBook)
https://doi.org/10.1007/978-3-031-43260-6

This Springer imprint is published by the registered company Springer Nature Switzerland AG
The registered company address is: Gewerbestrasse 11, 6330 Cham, Switzerland

Paper in this product is recyclable.

Acknowledgments

The development of this book spans multiple years of research. We extend our heartfelt gratitude to our collaborators and colleagues for all the inspiring discussions, for pointing us to new ideas, and for patiently contradicting us. Our writing journey began after teaching a summer course together with Willem Zuidema at the European Summer School for Logic, Language, and Information in 2021. We thank Willem for sharing his visionary ideas and for organizing the plenary discussion with the enthusiastic students that inspired this book.

Research is always a collaborative endeavor and our colleagues serve as an invaluable sounding board for shaping our thoughts. To name only a few collaborators, we would like to highlight the incredibly talented Ph.Ds. Charlotte Pouw and Ece Takmaz and our excellent co-authors Lena Jäger, Maria Barrett, and Stefanie Brandl. Your perspectives helped us better understand the many facets of cognitive plausibility.

Lisa deeply appreciates the privilege of being embedded in Amsterdam's vibrant NLP community. She expresses her gratitude to the researchers at the CLTL group at VU Amsterdam, and at the CLC and dialog modeling groups at the ILLC whose brilliance and dedication have been a constant source of inspiration. And she thanks her family for their patience because they matter most.

Nora is grateful for the stimulating working environment at the University of Copenhagen. She is especially thankful to her colleagues at the Centre for Language Technology, to Anders Søgaard for his guidance, and to the students of the MSc IT & Cognition program. She thanks the researchers at the Department of Computational Linguistics of the University of Zurich, a continuous pillar of support from the very beginning of her NLP journey. And, always, her husband Simon.

Graeme Hirst provided patient and motivating guidance through the process of writing our first book. We thank him and the anonymous reviewers who gave us extraordinarily constructive and encouraging feedback with great attention to detail.

Lastly, we genuinely hope that you enjoy reading this book as much as we enjoyed our collaboration over the past few years.

July 2023   Lisa Beinborn
                                                                  Nora Hollenstein

Contents

About the Authors

Lisa Beinborn is an assistant professor for Natural Language Processing at the Computational Linguistics and Text Mining Lab at Vrije Universiteit Amsterdam. Her research focuses on cognitively plausible language processing and cross-lingual models. She studied Computational Linguistics in Saarbrücken, Barcelona, and Bolzano, and obtained her Ph.D. in Computer Science at TU Darmstadt. She works on the interpretability of representational transfer and is interested in educational applications of NLP.

Nora Hollenstein is currently working at the Center for Language Technology of the University of Copenhagen and at the Department of Computational Linguistics of the University of Zurich. She obtained her Ph.D. from ETH Zurich working on cognitively inspired NLP. The focus of her research lies in improving and evaluating natural language processing applications with cognitive signals such as eye-tracking and brain activity recordings. She is especially interested in multilingual and multimodal NLP.

Introduction

Language is a powerful tool of human communication that provides elegant mechanisms for expressing highly complex phenomena. We use language every day and in all aspects of our lives. The versatility and variability of language make it a difficult subject for computational modeling, in contrast to more systematic sensor signals. Language follows underlying rules only to surprise us with exceptions and ambiguities on all linguistic levels and understanding its subtleties requires even more culture-specific knowledge than interpreting images.

We cannot derive an accurate static description of language because it dynamically evolves over time and across domains. More than 7,000 signed and spoken languages exist in the world covering a large spectrum of typological configurations [1, 2]. If we try to isolate a fundamental principle of language processing in scientific models, we will soon encounter a language with a complementary linguistic structure that creatively contradicts our assumptions.

In spite of these complexities, humans usually process language effortlessly. We are able to vary our language use to smoothly adapt to the target audience and dynamically integrate situational cues for seamless disambiguation. Natural language processing (NLP) research has already spent decades trying to understand how to computationally model language but complex reasoning tasks and creative constructions still lead to obvious failures of models. Nevertheless, the success of the field is undeniable. It attracts a continuously growing number of researchers and language processing models have become a key technology in our daily lives. These developments are strongly linked to the increasing availability of large amounts of training data and more efficient computing resources. Neural language models (LMs) are trained on terabytes of data and optimize millions of parameters to extract patterns from text.

When we train such a model to represent language, we fall back on a range of simplifying assumptions. For text-based models, we often expect that the text is formatted as one sen-

L. Beinborn and N. Hollenstein, *Cognitive Plausibility in Natural Language Processing*, Synthesis Lectures on Human Language Technologies, https://doi.org/10.1007/978-3-031-43260-6_1

tence per line, does not contain any images or special fonts, and is (mostly) typo-free. We implicitly assume that large text corpora are representative of modern language use and that optimizing the implemented language modeling objective relies on information that is relevant to understanding language. As researchers, we are usually aware that our assumptions oversimplify realistic scenarios. When we develop computational models, these assumptions become explicit in our modeling choices which makes it possible to directly object to them and empirically compare competing hypotheses. Building on the famous aphorism by Box [3] stating that "all models are wrong", Smaldino [4] discusses how "stupid models" can provide important insights into our misconceptions. He claims that working with computational models forces us to expose "our foolishness to the world" as a "price of seeking knowledge".

While the general benefit of computational modeling has been clearly established in multiple disciplines, we observe substantial disagreement in evaluating which properties characterize a "useful" model. Ruder et al. [5] describe how NLP research is only slowly evolving from purely performance-oriented experiments to integrating additional factors such as fairness, interpretability, computational efficiency, and multilingualism. Throughout this book, we propose integrating cognitive plausibility as an additional factor and agree with their call for more multi-dimensional research to capture interactions. We think that a multilingual perspective is required to learn more about cognitively plausible principles of language processing. And cognitively more plausible models can lead to computationally more efficient models.

Cognitive plausibility itself is a multi-faceted concept that varies considerably across disciplines. Computer scientists focus on the quantitative performance of the model, which should not be distinguishable from humans on static benchmark datasets. Neuroscientists focus on the biological plausibility of the model and aim to develop models of synaptic plasticity, which are evaluated on toy datasets that are much smaller than the common evaluation datasets in natural language processing. Psychologists focus on the plausibility of the learning processes in language models. They question the size and quality of the input data, examine memory and attention constraints, and explore learning curves. They try to isolate experimental factors by working with carefully designed stimuli that are often not representative of realistic language use.

We approach the concept of cognitive plausibility by diving into interpretability research and identifying the potential of its methods for cognitively inspired research questions. We focus on computational language models that are based on neural network architectures. These models are very powerful and their distributed approach is often motivated as being more cognitively plausible. Unfortunately, the expressivity of the model is gained at a loss of transparency. The high-dimensional matrix transformations that characterize neural models are hard to conceptualize for humans. When comparing models with millions of parameters, it becomes difficult to isolate the underlying modeling assumption that explains the differences. Interpretability methods are being developed to gain a better understanding of the inner workings and the inductive biases of neural models. These methods are often discussed from an engineering perspective focusing on quantifiable and measurable approaches to compare different models. We take a different stance: we explore interpretability methods through a cognitive lens by linking them to aspects of human language processing.

In this chapter, we first discuss why cognitive plausibility is a relevant factor for natural language processing research. We then define three dimensions of cognitive plausibility that we address in this book and provide an overview of how we approach their analysis.

1.1   Does Cognitive Plausibility Matter?

The cognitive plausibility of a model is an aspect that is commonly only implicitly targeted in natural language processing by assuming that higher average performance is obtained by better models and that better models are cognitively more plausible. We think that it is worthwhile to address cognitive plausibility more explicitly and want to initiate a more nuanced and in-depth discussion.

1.1.1   Human-Centered Natural Language Processing

In a recent data-driven survey on the values encoded in highly cited machine learning publications, the goals of developing a "human-like mechanism", "learning from humans" and being "transparent (to users)" can be found at the bottom of the list, while performance is the main driving force in 96% of the examined papers [6]. NLP research is strongly influenced by trends and developments in machine learning research but humans are central to our field: language is generated and developed by humans and language processing tools are directly used by humans (in contrast to neural models that are used for building machines or cultivating plants). Both, the input and the output of our models are characterized by human preferences, thus "human language processing is the ultimate gold standard for computational linguistics" [7].

Ethayarajh and Jurafsky [8] criticize that the progress in computational modeling is currently determined mostly by performance-oriented comparisons and propose to develop more user-centric leaderboards. They take a micro-economic approach to identify a model's utility for an end user and urge for more transparent reporting of practical factors such as model size, energy efficiency, and inference latency. In our opinion, the cognitive plausibility of a model is an underestimated factor of its utility.

The cognitive and economic aspects of model utility are not as unrelated as they may seem. Humans still outperform language models with respect to their linguistic generalization capabilities. We learn, understand, and produce language effortlessly, and while the cognitive mechanisms are not yet fully understood, it is clear that we do so very efficiently. As current computational models of language still struggle with phenomena that pose no problems for humans, it seems an obvious choice for us to turn to psycholinguistics and neurolinguistics for inspiration to build better computational models. If we want our models to acquire language as efficiently as humans and to generalize to new structures and contexts [9], we should reward cognitively more plausible architectures that simulate human transfer skills rather than models that excel in capturing statistical patterns of limited datasets [10].

1.1.2 Understanding Language Versus Building Tools

Natural language processing is an interdisciplinary research area by definition. It attracts researchers from a wide range of diverse backgrounds with very different research ambitions. When we develop computational models of language, two main goals can be distinguished. We might be driven by the vision to develop a computational model to better understand how humans process language. If we take a more practical perspective, we aspire to build a tool that automates language-related tasks to simplify our daily routines. While cognitive plausibility is clearly central to the first goal, it is less obvious for the tool-oriented approach.

One could argue that the cognitive plausibility of a model is irrelevant if the model works well enough for an intended use case. An information retrieval model that uses static keywords and templates can extract valuable information from scientific papers. While such a model can be useful for the average scenario, the user needs to take conscious countermeasures to account for its weaknesses (i.e., checking for alternative keywords and ascertaining that the extracted information is not framed within a negative context).

When working with neural language models, it becomes challenging to identify the patterns that determine the behavior of the model. Language input is complex and highly ambiguous and the model's decisions are learned by optimizing millions of parameters. With these two factors combined, it is almost impossible to compile clear descriptions for users in the form: if the input has property X, then the expected outcome quality is Y. The addition or omission of a single word might already flip the expected output label, but the model might not be sensitive to such changes if they have not been seen in the training data.

We are convinced that the ability to robustly anticipate the strengths and weaknesses of a model is a decisive factor in human-centered NLP. Cognitively more plausible models exhibit decision patterns that are more intuitive for human users because they are consistent with what they would expect from a colleague. We assume that cognitive consistency facilitates the practical application of a model because users can anticipate the reliability of its predictions and keep an eye on potential sources of error. For example, an essay scoring system that assigns a high grade to a well-written essay although it contains an argumentative fallacy is more plausible than a system that rewards an essay that is simply a bag of keywords. A tired human might have made the same mistake in the first case but could easily spot the word salad.

Language models have become the Swiss army knife of natural language processing because the finetuning methodology enables versatile adaptation to many tasks. High performance on a challenging subset of these tasks is often claimed to be an indicator of natural language understanding. Bender and Koller [11] claim that understanding requires a representation of meaning and initiated a discussion on whether meaning can be derived from form alone. They argue that meaning can only be interpreted with respect to the communicative intent of the utterance and define the conventional meaning of an expression as an abstraction over all possible contexts. As language models do not have access to functions of consciousness and self-awareness, they cannot capture forms of communicative intent. They can derive patterns from the training data but they cannot infer commonsense knowledge that is not explicitly expressed in textual sources due to the reporting bias [12]. The development of language models is advancing at a very fast pace and newer models master tasks that have been considered impossible to achieve. They can process and generate language to an extent that indicates a broad and growing set of linguistic competencies but we cannot conclude that they understand the meaning of words since they still fall short in many other respects such as grounding meaning in perception and action [13].

Our book presents a structured introduction to methods for analyzing the cognitive plausibility of language models. We do not aim to provide an absolute definition of the upper bound for a cognitively plausible model because we think that it is more useful to view cognitive plausibility as a graded concept.

1.1.3 Ethical Considerations

The release of ChatGPT [14] has led to a sudden recognition of language models reaching far outside academic target groups. With the wave of hastily released premature applications building on proprietary technology, the urgency of ethical scrutiny has become undeniable. Natural language processing is particularly sensitive to ethical problems because the way we use language to frame events, opinions, or feelings, affects our moral judgments and our perception of responsibility [15]. When language models are becoming increasingly cognitively plausible, we need to assure that they are not used in harmful ways. Ethical aspects should be considered at all levels of model implementation, starting with the data up to the misinterpretation of model outputs and the potential misuse of applications. In each chapter of this book, we discuss ethical aspects related to the respective methodology.

We discuss the problems that arise due to biases in the training data of language models and point out sustainability concerns with respect to computationally expensive training regimes. Many of the methods presented in this book rely on cognitive signals collected from human participants. When dealing with such sensitive data, it is imperative to adhere to privacy regulations. We discuss aspects of anonymization and overgeneralization and draw attention to systematic demographic biases. We expand on the problem of societal biases and discuss the trade-off between normative and descriptive ethics with respect to model behavior. We address the need for transparency about the limitations of our methodological choices and the importance of communication and traceability to ensure open pathways between academic research, society, and education.

1.2   Dimensions of Cognitive Plausibility

In cognitive science, model analyses are often characterized by distinguishing three levels proposed by Marr [16].[1] He described the computational level, the algorithmic level, and the implementational level of a model. The computational level targets the input–output patterns of the model and the operational constraints, the algorithmic level analyses the representations and procedures of the model, and the implementational level focuses on the hardware, i.e., the physical realization of the model. Over the years, the names for Marr's levels, their boundaries, and the interaction between them have been subject to many debates [17]. Nevertheless, they are still used today to characterize deep neural models [18].

In this book, we try to structure existing discussions of cognitive plausibility along three dimensions to facilitate the interpretation of cognitive claims about the strengths and weaknesses of language models. Our dimensions can be loosely linked to Maar's levels but the granularity differs. We first analyze the cognitive plausibility of behavioral patterns which can be linked to Marr's computational level.[2] We then further distinguish between representational structure and procedural strategies. This duality is also central in Marr's description of the algorithmic level.

In the field of natural language processing, Keller [7] proposed the development of cognitively plausible models of human language processing as "a challenge that requires the combination of research efforts in computational linguistics and psycholinguistics". They decomposed the task into a modeling challenge and a data and evaluation challenge.

Cognitive plausibility is often referenced when motivating choices for a language model architecture but the link to concrete cognitive functions usually remains rather shallow and metaphorical. We sketch a subset of choices for language model architectures in Chap. 2. For the remainder of the book, we abstract from the modeling challenge and focus more on the evaluation challenge. To this purpose, we introduce a range of human cognitive signals in Chap. 3. As we are focusing on neural models, the evaluation of the strengths and weaknesses of the model is closely linked to interpretability methods. We consider a model to be cognitively plausible if it makes similar decisions as humans (Chap. 4), uses a similar representational structure to encode knowledge as humans (Chap. 5), and applies similar procedural strategies as humans (Chap. 6).

1.2.1   Behavioral Patterns

The behavioral patterns of a model are characterized by its input–output pairs. As human labels are considered the gold standard for natural language processing, the goal of a good

---

[1] Marr's theory was published posthumously in 1982, but we only have access to the 2010 edition by MIT Press.

[2] The analysis framework that inspired Marr's distinction also referred to behavioral patterns [19].

model is to respond to a given input with a similar output as humans. Traditionally, this quality is assessed by looking at the average performance of a model on a given testset.

We argue that this is an oversimplification and discuss methods for instance-level evaluation. To analyze the cognitive plausibility of the model behavior, we need to identify more general behavioral patterns, for example, by examining subpopulations and designing diagnostic tests. We propose to integrate the concept of instance difficulty into the evaluation and link it to measures of model uncertainty. We assume that a model that fails to predict the outcome for difficult instances is more cognitively plausible than a model that fails on easy instances. We discuss how instance-level analysis, curriculum learning, and a multilingual perspective can contribute to cognitively more plausible models.

1.2.2   Representational Structure

Neural language models maintain internal representations of language input that are expressed as high-dimensional vectors. Human language processing performance is characterized by our ability to generalize to unseen compositions. Examining the type of knowledge that is encoded in the representational structure of the model makes it possible to analyze the generalization capabilities of neural models and identify potential over-reliance on memorization.

From a cognitive perspective, vector representations of language stimuli are difficult to interpret because we map symbolic representations into a continuous vector space. We provide an overview of methods to analyze the relative distribution in the vector space and discuss cognitively motivated probing approaches. We believe that cognitively plausible representations need to be situationally grounded and discuss representation learning approaches for multimodal and cognitive grounding.

1.2.3   Procedural Strategies

Static representational analyses of concepts, properties, and relations, do not scale well to the compositional power of contextualized models which learn to recombine input elements and integrate contextual information across layers. Novel analysis methods for determining relative importance and identifying selective attention can help us understand how models link information in a more dynamic fashion.

We discuss how local attribution and surprisal measures can be combined with cognitive signals to identify differences in the procedural strategies between humans and models. In order to build models with cognitively more plausible generalization skills, multi-task learning and transfer learning seem to be promising approaches. We also discuss how the inductive bias of a model can be modified using linguistic or cognitive information.

1.3   Analyzing Cognitive Plausibility

This introductory book is based on the idea that we can learn more about the cognitive plausibility of computational language models by integrating signals of cognitive processing load in humans into interpretability analyses. We are targeting primarily advanced master's students and early doctoral researchers investigating topics at the intersection of natural language processing, machine learning, and cognitive science. We hope that they can use the chapters as a gateway to the addressed issues and aim at providing an explanatory starting point with an initial set of useful references for further research.

In Chap. 2, we first introduce the required background knowledge about language modeling. Our approach to cognitive plausibility heavily relies on information that can be drawn from cognitive signals. We introduce a range of cognitive signals that reveal patterns of language processing in humans and discuss methodological characteristics in Chap. 3. The cognitive signals we can collect differ with respect to temporal accuracy, the degree of consciousness involved, and the levels of linguistic processing they reflect. We showcase how the research question affects the choice of the most suitable cognitive signal. For example, gaze patterns can inform us about cognitively plausible attention mechanisms and brain activation signals can provide a window into the processes of syntactic disambiguation. We try to incorporate a wide range of phenomena while focusing on text processing and language comprehension.

We then describe the details of the three dimensions of cognitive plausibility introduced above in Chaps. 4–6. In each of these chapters, we first introduce methods for extracting information from the model. These methods are heavily inspired by the field of interpretability and explainability research. We then explain how these methods can be combined with cognitive signals to compare model patterns to human language processing. Cognitive plausibility is a vast interdisciplinary field and our book raises more questions than it answers. We discuss our main recommendations for studying cognitive plausibility and identify promising directions for future research in Chap. 7.

References

1. David Eberhard, Gary Simons, and Charles Fennig (eds.). *Ethnologue: Languages of the World*. SIL International, Dallas, Texas, twenty-fifth edition, 2022. http://www.ethnologue.com.
2. Matthew S. Dryer and Martin Haspelmath, editors. *WALS Online*. Max Planck Institute for Evolutionary Anthropology, Leipzig, 2013. https://wals.info/.
3. George E. P. Box. Science and statistics. *Journal of the American Statistical Association*, 71(356):791–799, 1976. https://doi.org/10.1080/01621459.1976.10480949. https://www.tandfonline.com/doi/abs/10.1080/01621459.1976.10480949.
4. Paul E Smaldino. Models are stupid, and we need more of them. In *Computational social psychology*, pages 311–331. Routledge, 2017.
5. Sebastian Ruder, Ivan Vulić, and Anders Søgaard. Square one bias in NLP: Towards a multidimensional exploration of the research manifold. In *Findings of the Association for Com-*

putational Linguistics: ACL 2022, pages 2340–2354, Dublin, Ireland, May 2022. Association for Computational Linguistics. https://doi.org/10.18653/v1/2022.findings-acl.184. https://aclanthology.org/2022.findings-acl.184.

6. Abeba Birhane, Pratyusha Kalluri, Dallas Card, William Agnew, Ravit Dotan, and Michelle Bao. The values encoded in machine learning research. In *2022 ACM Conference on Fairness, Accountability, and Transparency*, pages 173–184, 2022.

7. Frank Keller. Cognitively plausible models of human language processing. In *Proceedings of the 48th Annual Meeting of the Association for Computational Linguistics: Short Papers*, pages 60–67, 2010.

8. Kawin Ethayarajh and Dan Jurafsky. Utility is in the eye of the user: A critique of NLP leaderboards. In *Proceedings of the 2020 Conference on Empirical Methods in Natural Language Processing (EMNLP)*, pages 4846–4853, Online, November 2020. Association for Computational Linguistics. https://doi.org/10.18653/v1/2020.emnlp-main.393. https://aclanthology.org/2020.emnlp-main.393.

9. Dani Yogatama, Cyprien de Masson d'Autume, Jerome Connor, Tomas Kocisky, Mike Chrzanowski, Lingpeng Kong, Angeliki Lazaridou, Wang Ling, Lei Yu, Chris Dyer, et al. Learning and evaluating general linguistic intelligence.*arXiv preprint* arXiv:1901.11373, 2019.

10. Tal Linzen. How can we accelerate progress towards human-like linguistic generalization? In *Proceedings of the 58th Annual Meeting of the Association for Computational Linguistics*, pages 5210–5217, Online, July 2020. Association for Computational Linguistics. https://doi.org/10.18653/v1/2020.acl-main.465. https://aclanthology.org/2020.acl-main.465.

11. Emily M. Bender and Alexander Koller. Climbing towards NLU: On meaning, form, and understanding in the age of data. In *Proceedings of the 58th Annual Meeting of the Association for Computational Linguistics*, pages 5185–5198, Online, July 2020. Association for Computational Linguistics. https://doi.org/10.18653/v1/2020.acl-main.463. https://aclanthology.org/2020.acl-main.463.

12. Vered Shwartz and Yejin Choi. Do neural language models overcome reporting bias? In *Proceedings of the 28th International Conference on Computational Linguistics*, pages 6863–6870, Barcelona, Spain (Online), December 2020. International Committee on Computational Linguistics. https://doi.org/10.18653/v1/2020.coling-main.605. https://aclanthology.org/2020.coling-main.605.

13. Brenden M Lake and Gregory L Murphy. Word meaning in minds and machines. *Psychological review*, 2021.

14. Open AI. Introducing chatgpt, 2022. https://openai.com/blog/chatgpt.

15. Gosse Minnema, Sara Gemelli, Chiara Zanchi, Tommaso Caselli, and Malvina Nissim. Dead or murdered? predicting responsibility perception in femicide news reports. In *Proceedings of the 2nd Conference of the Asia-Pacific Chapter of the Association for Computational Linguistics and the 12th International Joint Conference on Natural Language Processing (Volume 1: Long Papers)*, pages 1078–1090, Online only, November 2022. Association for Computational Linguistics. https://aclanthology.org/2022.aacl-main.79.

16. David Marr.*Vision: A computational investigation into the human representation and processing of visual information*. 2010.

17. Valerie G. Hardcastle and Kiah Hardcastle. Marr's levels revisited: Understanding how brains break. *Topics in Cognitive Science*, 7(2):259–273, 2015. https://doi.org/10.1111/tops.12130. https://onlinelibrary.wiley.com/doi/abs/10.1111/tops.12130.

18. Jessica Hamrick and Shakir Mohamed. Levels of analysis for machine learning, 2020. https://arxiv.org/abs/2004.05107.

19. Tomaso Poggio. Afterword: Marr's Vision and Computational Neuroscience. In *Vision: A Computational Investigation into the Human Representation and Processing of Visual Information*. The MIT Press, 07 2010. ISBN 9780262514620. https://doi.org/10.7551/mitpress/9780262514620.003.0009.

Foundations of Language Modeling

2

Representing the subtleties of natural language in a way that can be computationally processed is a fundamental challenge of natural language processing. Language is developed and shaped by humans and its interpretation is based on arbitrary social conventions. We cannot develop precise mappings from sentences to meaning because language use is often ambiguous and vague. It varies depending on the situational context, and it dynamically evolves over time. In natural language processing, computational models of language are applied to automatically process and analyze written or spoken language data.

Traditional computational language models were developed to determine the probability of a sentence [1]. They predict the most probable next token in a sequence based on co-occurrence patterns in the training corpora. Language models were used to rank candidate sequences in speech recognition [2] or machine translation [3] as part of noisy channel models. The distributional representations calculated by language models turned out to be a good basis for many other natural language processing tasks, for example, for topic modeling [4] or grammar correction [5].

The language modeling objective can be optimized using neural language models [6]. By drastically increasing the size of the training corpora and using variations of the training objective [7–9], pretrained neural language models have quickly become the best-performing standard for many natural language processing tasks [10]. These large pretrained models can be finetuned to different tasks and domains with relatively little data and computational power and often perform on par with human experts [11]. In this chapter, we describe the fundamental methodological concepts and modeling decisions behind these models.

© The Author(s), under exclusive license to Springer Nature Switzerland AG 2024   11
L. Beinborn and N. Hollenstein, *Cognitive Plausibility in Natural Language Processing*, Synthesis Lectures on Human Language Technologies,
https://doi.org/10.1007/978-3-031-43260-6_2

2.1   Methodological Concepts

We provide a very brief introduction to language modeling to Synonym the most relevant terms for this book. More details can be found in Chap. 10 of Goodfellow et al. [12] and Chaps. 3 and 9 of Jurafsky and Martin [13]. Language models were originally optimized to predict the next token based on a sequence of given tokens. In statistical approaches, a language model calculates the conditional probability that the example sequence *Only time will* is followed by the token *tell*.

$$P(\text{tell}|\text{Only time will}) = \frac{P(\text{Only time will tell})}{P(\text{Only time will})} \tag{2.1}$$

More formally, in generative tasks, the goal is to find the token w_n that maximizes the probability of the sequence $w_1...w_n$:

$$\arg \max_{w_n} P(w_n|w_1 \, w_2 \, ... w_{n-1}) \tag{2.2}$$

As many sequences cannot be observed in the training data, n-gram based language models simplify this probability estimation by shortening the context to the $n - 1$ previous tokens. The probabilities of the n-gram sequences are then combined using the Markov Assumption and smoothing techniques (see [14] for more details).

2.1.1   Language Modeling with Recurrent Neural Networks

The language modeling objective can be optimized by a neural model using a recurrent architecture [6]. In this case, the sequence is fed into the model token by token. At each step, the model outputs a probability distribution over the vocabulary.

For a basic recurrent neural network, the input tokens are represented as one-hot vectors $\mathbf{x_i}$ which are multiplied by an embedding matrix $\mathbf{W}$. To obtain the hidden representation $\mathbf{h_i}$, we multiply a weight matrix $\mathbf{V}$ with the hidden representation of the previous token $\mathbf{h}_{i-1}$ and apply a non-linear function σ.[1]

$$\mathbf{h_i} = \sigma(\mathbf{Vh}_{i-1} + \mathbf{Wx_i}) \tag{2.3}$$

$$\mathbf{o_i} = \mathbf{Uh_i} \tag{2.4}$$

$$\hat{\mathbf{y}}_i = \text{softmax}(\mathbf{o_i}) \tag{2.5}$$

The output representation $\mathbf{o_i}$ is calculated by multiplying the weight matrix $\mathbf{U}$ with the hidden representation.[2] The number of dimensions of $\mathbf{o_i}$ is equivalent to the number of tokens in the

---

[1] For readability, we drop the bias in the notation as is common in related work.
[2] Sometimes $\mathbf{W}$ and $\mathbf{U}$ are the same matrix.

Fig. 2.1 Language modeling in a recurrent neural network

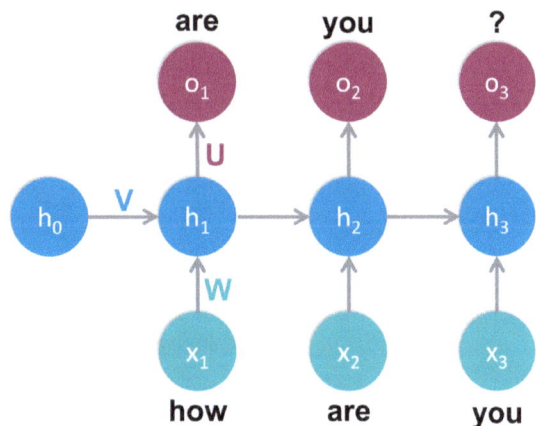

vocabulary of the model. To predict the next token $\mathbf{x_{i+1}}$, the softmax function is applied to the output activations to obtain a probability distribution over the vocabulary, and the token with the maximum probability is selected (Fig. 2.1).

The weight matrices $\mathbf{W}$, $\mathbf{V}$, and $\mathbf{U}$ are initialized randomly. During model training, they are optimized on a large corpus of texts. The cross-entropy loss is calculated by comparing the predicted probability distribution for the next token to the observed token in the data.

Language modeling is a task that exhibits path dependency. For example, if the model predicted a token with a different part-of-speech tag (*Only time will eventually*), all following predictions would be based on a different syntactic interpretation of the context. The alternative productions might still be valid sequences of the language but language models are optimized to represent the training data as closely as possible. To reduce path dependency, a mechanism called teacher forcing is applied during training. When predicting the next token, the (possibly wrong) prediction by the model for the previous token is ignored and the observed sequence in the corpus is used as context instead.

2.1.2   Evaluating Language Models

A trained language model can assign a probability p to new sequences $w_1, w_2, ..., w_i,$ $w_{i+1}, ..., w_n$. For each word w_i, we can calculate the surprisal s_i of the model of seeing this word given the previous context by taking the negative log probability of observing the sequence:

$$\text{Probability } p(w_i) = p(w_i | w_1, ..., w_{i-1}) \tag{2.6}$$

$$\text{Surprisal } s_i = -log\ p(w_i) \tag{2.7}$$

Surprisal is often used to analyze cognitive signals, for example, to predict reading times or N400 effects in EEG data [15–17]. Surprisal of individual words can be summarized as the entropy H by taking the weighted average over all tokens W in the corpus [18]. Language models are commonly evaluated by the perplexity metric PPL, which projects the entropy from log space back to linear space and expresses how well the model can predict the training data.

$$\text{Entropy } H(W) = \sum_{w_i \in W} p(w_i)s_i \qquad (2.8)$$

$$\text{Perplexity } PPL(W) = 2^{H(W)} \qquad (2.9)$$

If a model assigns a low probability to a sequence, its perplexity is high. Perplexity can be calculated over the whole training corpus or as an average over perplexity scores for each sentence, depending on the characteristics of the data.

2.1.3 Language Modeling as Representation Learning

Neural models are more likely to converge if the vector representations of the input are initialized meaningfully, i.e., the initial weights are already close to a good solution [19]. The performance of computer vision models improved drastically when they were initialized using a model pre-trained for image classification [20]. In natural language processing, language modeling has been found to be a suitable pre-training objective when it is optimized on sufficiently large datasets [21]. Using a language modeling objective has two advantages: 1) It does not require additional human annotations because the gold label (the word to be predicted) is provided by the text itself which was written by humans. This is called a self-supervised learning setup. 2) Optimizing for the language modeling objective requires both semantic and syntactic knowledge [22]. The predicted next word needs to fit into the selectional preferences, e.g., if *She drinks a* ... is completed with a noun, we expect it to be something drinkable like *beer* or *coffee* or a container of a drink like *bottle* or *cup*. The word also needs to fit the syntactic constraints: the plural *coffees* would be ungrammatical in this example.

Pretrained language models now serve as a valuable starting point for many natural language processing tasks. For the remainder of this book, we usually have a large, pretrained language model in mind when discussing the cognitive plausibility of models. They can be customized for different scenarios using prompting, finetuning, or adapters.

Prompting
Prompting is a technique that uses the ability of a language model to fill in blanks and assign probabilities to its generated tokens [23]. Instead of using the original pretraining objective directly, prompt engineering is used to convert a classification objective into a

mask-fill task. For example, when the objective is to predict the sentiment of a sentence such as *I can't listen to this song anymore*, one would transform it into a prompt by combining it with a template such as *Overall, it is simply a [X] song*. If the language model probabilities for predetermined negative fillers such as *terrible, irritating*, or *horrendous* are higher than for respective positive fillers such as *beautiful* or *fantastic*, the answer is mapped into a negative label, which is used as the classification output. Prompting exploits the vast pretraining knowledge of the model and is particularly useful for zero-shot or few-shot classification scenarios, i.e., when no or relatively few training instances are available. The technique of comparing language model probabilities is often used when analyzing the cognitive plausibility of representational structure (see Chap. 5). Prompting has become a very promising research direction as it requires fewer computational resources [24] but it comes with its own challenges. It can be tricky to find the right strategies for designing useful templates, for searching and consolidating the answers, and for mapping answers back to labels. When trained with so-called chain-of-thought prompting, language models can even be instructed to generate potential explanations for their answer [25]. The prompting description we provide here is restricted to tuning-free prompting. For higher classification performance, prompting is often combined with finetuning.

Finetuning

Finetuning is a generic term that can refer to a wide range of techniques for updating the parameters of a pretrained language model to adapt it to a new task or domain. When it was introduced [21], it referred to a supervised classification setting for which an additional classification layer is added on top of a language model. The entire model is then finetuned to optimize a new loss function specific to the task. For each instance, it generates a probability distribution over the output labels. The idea of finetuning has now been extended to a wide range of scenarios including, for example, continual pretraining for domain adaptation [26] or prompt tuning [27].[3] In all these scenarios, gradient updates are calculated through the entire model, which is computationally expensive and requires a substantial amount of supervised data.

Adapters

A more efficient alternative to full finetuning are adapter modules which are inserted in between the transformer layers of a pretrained language model [29, 30]. During finetuning, only the small subset of adapter weights are optimized and the original model weights remain fixed. [31] provide a large library of trained adapters and the modular architecture makes it possible to quickly exchange them to enable knowledge transfer of a model to different tasks, domains, and languages.

---

[3] It should be noted that a model might lose its pretraining capabilities due to finetuning; a phenomenon known as catastrophic forgetting [28], see Sect. 6.3.2.

2.2   Modeling Decisions

Language models are usually evaluated quantitatively to compare the efficiency of different modeling decisions. Unfortunately, model training is so expensive that models differ in a multitude of experimental parameters which makes it difficult to isolate independent variables that are responsible for performance gains. In this section, we describe how different experimental parameters for language modeling can be linked to aspects of cognitive plausibility. At the time of writing the main content for this book, the model RoBERTa was the state-of-the-art for language modeling [32]. As the field is developing at an overwhelming speed, many new models have been introduced since then. We try to abstract from individual models and focus on the more general underlying questions.

2.2.1   Target Objective

The representations that a language model learns are directly influenced by the target objective that it optimizes during training. Predicting the next token is the classical training objective for language models that process the input incrementally from left to right [33]. This incremental constraint is quite restrictive as ambiguous sequences can often only be resolved after seeing the full context. For example, when processing the sequence *The old train*, the most probable expectation would be that *train* is a noun that refers to a railed vehicle. If the sequence continues with *The old train the young in our volleyball club*, we need to reinterpret the meaning of *train* to be a verb referring to the activity of teaching or instructing someone. In order to strengthen the contextualization of language model representations, Devlin et al. [9] introduced the masked language modeling (MLM) objective which is inspired by educational cloze tests [22]. In this scenario, the model can process the full input sequence except for a small ratio of tokens that have been replaced with the MASK token. For each masked token, the model predicts a probability distribution over the vocabulary. In our example, the input would be *The old MASK the young in our volleyball club* and ideally, the model would predict fitting verbs such as *feed, protect, train, sanction*.

When applying the masked language modeling objective in practice, usually more than one token per sentence is masked to make the model more robust. This makes the prediction task harder as the model cannot account for dependencies between the masked tokens. Masking input tokens and neglecting the dependencies between the masked positions can therefore be seen as *corrupting* the input with masks. In language learning, such a corruption approach is known as a reduced redundancy test [34]. Fluent speakers are quite robust to reduced redundancy (for example, when having a phone call in a noisy environment) and can even keep track of long-range dependencies which remains a challenge for language models [35].

To reduce the dependency between masked tokens, Yang et al. [36] propose permutation language modeling, which maximizes the expected likelihood for *all* possible permutations of the input. The context for each token position consists of varying combinations of tokens from both left and right such that the model learns to use contextual information from all positions independent of sequential constraints. This approach is motivated by engineering considerations and cannot be mapped to cognitively plausible processing mechanisms as humans do not assign equal importance to all tokens to anticipate what comes next [37].

The incrementality of language processing has been widely discussed in language comprehension research. During listening, humans generally process language incrementally and make predictions about upcoming events. The ability to maintain parallel hypotheses and to backtrack to repair an erroneous interpretation is a fundamental skill for finegrained language comprehension. Eye-tracking technology intuitively visualizes such techniques of anticipation and repairs during more conscious reading comprehension [38–40]. Saccades in the fixation path indicate that preceding tokens are read again to repair or confirm the initial interpretation. The original next-word-prediction objective can be seen as an online processing approach comparable to the first pass in reading [15], whereas masked language modeling rather captures the representation after repairing and re-interpreting the full sequence.

Language is compositional and humans use it in creative ways. State-of-the-art language models are trained on millions of sequences and we can still easily come up with a new sequence that the model has never seen such as *"My cockatoo is getting nervous because my neighbor laughs about her tax returns"*. The robustness of a model to novel sequences can be increased by introducing a small portion of noise to the input, for example, by randomly replacing tokens with other tokens [9]. Clark et al. [41] present a related pretraining approach called Electric that does not rely on the masking procedure. Instead, it uses contrastive noise estimation which is comparable to negative sampling for static word embeddings. For every token position, the model learns to distinguish between the originally fitting token and a set of unlikely tokens in this context (without having to predict a probability distribution over the entire vocabulary).

In order to go beyond representing tokens in the local context, many language models combine the token-level target objective with a sequence-level objective which evaluates the relation that holds for a pair of input sequences A and B. The most popular sequence-level relations are predicting whether the second sequence B is likely to be found directly after the first sequence A in a text (Next Sentence Prediction) [9]. The model is optimized by mixing positive examples of correct consecutive sequences (A, B) with negative examples in which A is paired with a random sequence X from the dataset. A semantically more finegrained objective is the prediction of entailment [42]. In this scenario, the model predicts the relationship between the first sequence A (the premise) and the second sequence B (the hypothesis). B entails A if it logically follows from A, e.g., *Theo is singing with his daughter* entails that *Theo has a child*.

Table 2.1 Word segmentation examples for (1) Chinese and (2) Finnish

(1)	我	喜欢	看	电视。
	I	*like*	*to watch*	*TV.*
(2)	Iltapäivällä	menen	kauppaan.	
	In the afternoon	*I go*	*to the store.*	

2.2.2 Input Units

Language can be interpreted as a stream of information: texts are sequences of characters and speech is a sequence of sounds. When we process language, we segment the input stream into smaller meaningful units. In natural language processing, we refer to this process as tokenization. As our field has been dominated by the development of English approaches [43], most earlier models and resources interpret words as the central unit. For many other languages, the word-based approach is not ideal. Chinese is segmented by combinations of characters and word boundaries cannot be approximated by white-space boundaries (as in 我喜欢看电视。, see segmentation example (1) in Table 2.1). Agglutinative languages are morpheme-based and tokens are composed of multiple meaning-bearing units. Example (2) in Table 2.1 is expressed using eight tokens in English but only three in Finnish.

Most word-based language models allocate a fixed vocabulary during training whereas humans can creatively come up with novel words using generative processes. For example, the neologism *workation* gained popularity during the pandemic. It blends the tokens *work* and *vacation* to refer to a remote working setup that combines job obligations with leisure. In order to account for unknown words, the preferred choice for the input unit has recently moved from words to so-called subwords.

Algorithms for subword tokenization segment the input into patterns of frequently co-occurring characters. Using smaller input units makes it possible to address the 'unknown word problem' and to transfer common roots across languages in multilingual models. However, the purely statistical segmentation approach often leads to cognitively implausible splits. For example, Fekete [44] finds that the tokenizer for multilingual BERT, interprets *screen* as a single token but splits *sunscreen* into *sun ##sc ##reen*. The word *glasses* is split into linguistically meaningful morphemes *glass ##es*, but *sunglasses* is greedily tokenized as *sung ##lasse ##s* which does not properly reflect the underlying linguistic structure.

Subword and character information are crucial in state-of-the-art neural language models [45, 46] because they can improve their generalization and robustness capabilities [47, 48]. Character-level features are beneficial for morphologically rich and low-resource languages [49, 50]. Another advantage of purely character-based language models is that they have a much smaller vocabulary, i.e., the total number of characters. Such models are often used for Chinese, Japanese, and Korean models [51, 52]. For these languages, it is also common to include sub-character information in language models due to the compositional structure of their orthographical system.

Subcharacter-level information can enhance the models by providing background knowledge of the underlying structures of words in a language [53]. For instance, Korean language models are often trained on *Jamos* (phoneme-like units), the smallest unit of the Korean script [54, 55]. This reduces the vocabulary size and injects syntactic and semantic information into the model that is difficult to access with conventional character- or token-level units [56]. Additionally, including sub-character information can reduce the size of the training data. For example, a Korean BERT model using sub-character information requires less training data than previous sub-token models [57].

The best choice for the input unit is still an unresolved question in language processing whereas computer vision seems to have settled on pixels. Recent work also claims that pixel-based segmentation could be an option for language processing to increase robustness [58]. This approach would make our models sensitive to the font in which a text is represented.

, Not only the smallest unit of segmentation is a debatable choice for language processing but also the boundaries for input sequences. Language modeling approaches differ in whether they allow individual sequences (usually sentences) as input, pairs of sequences, or entire paragraphs or documents. Shorter input sequences are usually padded to a fixed maximum length.

2.2.3 Processing Order

An important constraint that has been formulated for cognitively plausible language processing models is incrementality.[4] We extract as much syntactic and semantic information as possible from each token as we see it and integrate it into our current interpretation [59, 60]. Traditional n-gram-based language models follow this paradigm and process the input incrementally (from left to right for languages with a left-to-right writing system). In neural language models, sequential processing can be enforced by using a recurrent architecture. Recurrent neural networks incrementally build up a representation of the input one token at a time. At each step, the representation of all previous input tokens is integrated into the current representation. In theory, recurrent neural networks are thus able to integrate infinitely long contexts to resolve long-distance dependencies.[5] However, not all input tokens are equally important for predicting the next word. For example, in the question *Are you going to have a hot tall decaf half-soy white chocolate mocha with me?* we do not need to keep track of all the adjectives to predict the preposition *with* but the tokens *you* and *have* provide important information. In order to model selective attention, gating mechanisms have been introduced. The gates are in principle additional weight matrices that operate sequentially to indicate how much information is propagated. The most popular variant of gated recurrent

---

[4] Also referred to as immediacy.

[5] In practice, it has been shown that this is rarely the case [61].

networks are long short-term memory networks [62]. These networks combine three gates (forget, add, output) to modulate information. The term "short-term memory" reminds us of the processing patterns by humans and the distinction between long-term and short-term memory. However, the terms are used here rather as a metaphor to conceptually explain the functionality of matrix operations.

The directional models are computationally expensive because the recurrent connections impede parallelization of the calculation of the hidden states [63]. Transformer models make it possible to integrate sequential contexts without imposing an order on the calculation of the hidden states. The order of the tokens is only indicated by positional embeddings. They represent the relative position of a token within the sequence but the representation of each token can be calculated independently [9]. The relations between tokens are modeled by a mechanism called self-attention. For the calculation of self-attention, no sequential constraint is incorporated: each word in a sequence can attend to every other word in the same sequence. Self-attention is implemented by multiplying the input vector x_i of an input token with three additional weight matrices W^Q, W^K, and W^V (query, key, and value) that are optimized during training.

$$\text{query: } q_i = W^Q x_i \tag{2.10}$$

$$\text{key: } k_i = W^K x_i \tag{2.11}$$

$$\text{value: } v_i = W^V x_i \tag{2.12}$$

Each possible combination of input tokens x_i and x_j is assigned a score that is the product of the query representation of token x_i and the key representation of token x_j normalized by the square root of the number of dimensions d_k. The attention vector α_{ij} is then determined by taking the softmax over all token scores $s(x_i, x_k)$ to determine the proportional relevance of token x_j for token x_i.

$$\text{score: } s(x_i, x_j) = \frac{q_i \cdot k_j}{\sqrt{d_k}} \tag{2.13}$$

$$\text{attention: } \alpha_{ij} = \frac{\exp(s(x_i, x_j))}{\sum_k \exp(s(x_i, x_k))} \tag{2.14}$$

$$\text{output: } y_i = \sum_j \alpha_{ij} v_j \tag{2.15}$$

In human processing, attention depends on the perspective. We can direct our attention to different aspects of the input [64]. Transformer models commonly learn multiple attention heads, i.e., multiple sets of query, key, and value matrices. It is assumed that the model then learns to use the heads to attend to the input from different language processing perspectives, for example, local semantics, grammatical agreement, and co-reference resolution. Self-attention has emerged as an engineering solution to process larger amounts of training data in a parallel fashion. The attention heads have an important role in the good performance of transformers but the role of self-attention as a processing strategy is not yet clear.

2.3   Ethical Aspects

Language models have become the most powerful tool in natural language processing. They form the key technology for applications that we use every day such as semantic search, text completion, or machine translation. Our students cannot imagine not having access to these tools anymore. With power comes societal responsibility because every technology can be abused. Language models can generate fake reviews to create economic imbalances or automate social media comments to influence the political landscape. These examples illustrate forms of explicit abuse with criminal intentions, but the application of language models can also have unintentional consequences, for example by propagating implicit biases of the model into decision processes.

Language models learn patterns from training data which means that they also acquire a sensitivity to the societal biases inherent in the training data. Recent analyses find a large range of systematic biases including racism [65], sexism [66], religion [67], and bias against people with disabilities [68]. Identifying and controlling these biases is not an easy task. Antoniak and Mimno [69] find that bias measurement methods are often affected by the individual preferences of the researchers involved and Gonen and Goldberg [70] show that current debiasing methods are insufficient. When left unnoticed, harmful biases in the training data end up in downstream applications and reinforce existing socio-economic gaps. For example, Koenecke et al. [71] show that speech recognition systems by large tech companies perform substantially worse for African-American participants. This leads to disadvantages when the technology is applied in virtual assistants or for dictation purposes in professional environments.

As these discriminatory consequences mostly affect minorities who are not well-represented in developer teams and decision-making units, they often remain unnoticed [72]. Words and phrases that are explicitly derogatory can be detected by applying domain-specific lexical filters [73], but more indirect harmful expressions such as online abuse and subtle forms of negativity in microaggressions are much harder to capture systematically [74, 75].

Transformer-based language models only develop generalization abilities when pre-trained on large amounts of data [76]. The required computing resources for such large-scale training regimes are economically and ecologically unsustainable and only available to a selected few. Strubell et al. [77] quantify the computational resources required for a language modeling experiment as accumulating to 27 years of computing on a single GPU which averages to "about 60 GPUs running constantly throughout the 6-month duration of the project". They recommend that training time and sensitivity to hyperparameters should be taken into consideration when comparing models to encourage more computationally efficient hardware and algorithms. We expect that cognitively more plausible models will be able to learn and generalize more efficiently from smaller amounts of data.

The role of the size of language input is still an open question in language acquisition research because of the interdependencies with other factors of the language learning environment [78–80]. It is difficult to assess the number of words that a child is exposed to during learning but even generous estimates are no comparison to the enormous training data required by state-of-the-art language models [81]. Recent analyses indicate that language models can reliably encode syntactic and semantic features already with several million words in the training data [82] but much larger datasets are required to acquire commonsense knowledge to enable generalization in typical evaluation tasks [83]. These findings indicate a trade-off between optimizing for models with encyclopedic knowledge and cognitively plausible models.

The size of the training data alone does not guarantee a reflection of diversity [75]. Even large web-crawled datasets can be surprisingly narrow in the topics, text genres, and populations they represent, depending on the crawling depth and methodology. The stylistic patterns that language models learn are commonly shaped by relatively small communities that are over-represented in online data. Instead of aiming solely for scale and leaving bias mitigation as a post-hoc challenge, ethical considerations need to be proactively integrated already in the early stages of model development [84]. The EU guidelines for trustworthy AI state that for the principle of fairness, avoidance of unfair bias is an essential building block for ensuring fairness and should be addressed both in the data collection stage and during model development [85]. Current approaches for avoiding implicit biases focus on reverse-engineered countermeasures to filter and modify the model output. A more fruitful measure would be to avoid the amplification of hateful content through more careful curation and transparent documentation of the training data [86].

References

1. Claude E Shannon. A mathematical theory of communication. *The Bell system technical journal*, 27(3):379–423, 1948.
2. Marcello Federico, Nicola Bertoldi, and Mauro Cettolo. Irstlm: an open source toolkit for handling large scale language models. In *Ninth Annual Conference of the International Speech Communication Association*, 2008.
3. Holger Schwenk and Philipp Koehn. Large and diverse language models for statistical machine translation. In *Proceedings of the Third International Joint Conference on Natural Language Processing: Volume-II*, 2008. https://aclanthology.org/I08-2089.
4. Hanna M Wallach. Topic modeling: beyond bag-of-words. In *Proceedings of the 23rd international conference on Machine learning*, pages 977–984, 2006.
5. Y. Albert Park and Roger Levy. Automated whole sentence grammar correction using a noisy channel model. In *Proceedings of the 49th Annual Meeting of the Association for Computational Linguistics: Human Language Technologies*, pages 934–944, Portland, Oregon, USA, June 2011. Association for Computational Linguistics. https://aclanthology.org/P11-1094.

6. Tomas Mikolov, Martin Karafiát, Lukas Burget, Jan Cernockỳ, and Sanjeev Khudanpur. Recurrent neural network based language model. In *Interspeech*, volume 2, pages 1045–1048. Makuhari, 2010.

7. Tom Brown, Benjamin Mann, Nick Ryder, Melanie Subbiah, Jared D Kaplan, Prafulla Dhariwal, Arvind Neelakantan, Pranav Shyam, Girish Sastry, Amanda Askell, et al. Language models are few-shot learners. *Advances in neural information processing systems*, 33:1877–1901, 2020.

8. Matthew E. Peters, Mark Neumann, Mohit Iyyer, Matt Gardner, Christopher Clark, Kenton Lee, and Luke Zettlemoyer. Deep contextualized word representations. In *Proceedings of the 2018 Conference of the North American Chapter of the Association for Computational Linguistics: Human Language Technologies, Volume 1 (Long Papers)*, pages 2227–2237, New Orleans, Louisiana, June 2018. Association for Computational Linguistics. https://doi.org/10.18653/v1/N18-1202. https://aclanthology.org/N18-1202.

9. Jacob Devlin, Ming-Wei Chang, Kenton Lee, and Kristina Toutanova. BERT: Pre-training of deep bidirectional transformers for language understanding. In *Proceedings of the 2019 Conference of the North American Chapter of the Association for Computational Linguistics: Human Language Technologies, Volume 1 (Long and Short Papers)*, pages 4171–4186, Minneapolis, Minnesota, June 2019. Association for Computational Linguistics. https://doi.org/10.18653/v1/N19-1423. https://aclanthology.org/N19-1423.

10. Alex Wang, Amanpreet Singh, Julian Michael, Felix Hill, Omer Levy, and Samuel Bowman. GLUE: A multi-task benchmark and analysis platform for natural language understanding. In *Proceedings of the 2018 EMNLP Workshop BlackboxNLP: Analyzing and Interpreting Neural Networks for NLP*, pages 353–355, Brussels, Belgium, November 2018. Association for Computational Linguistics. https://doi.org/10.18653/v1/W18-5446. https://aclanthology.org/W18-5446.

11. Linting Xue, Noah Constant, Adam Roberts, Mihir Kale, Rami Al-Rfou, Aditya Siddhant, Aditya Barua, and Colin Raffel. mT5: A massively multilingual pre-trained text-to-text transformer. In *Proceedings of the 2021 Conference of the North American Chapter of the Association for Computational Linguistics: Human Language Technologies*, pages 483–498, Online, June 2021. Association for Computational Linguistics. https://doi.org/10.18653/v1/2021.naacl-main.41. https://aclanthology.org/2021.naacl-main.41.

12. Ian Goodfellow, Yoshua Bengio, and Aaron Courville. *Deep Learning*. MIT Press, 2016. http://www.deeplearningbook.org.

13. Daniel Jurafsky and James H Martin. Speech and language processing, 3rd edition. 2000.

14. Andreas Stolcke. Srilm-an extensible language modeling toolkit. In *Seventh international conference on spoken language processing*, 2002.

15. Vera Demberg and Frank Keller. Data from eye-tracking corpora as evidence for theories of syntactic processing complexity. *Cognition*, 109(2):193–210, 2008.

16. Michael Hahn and Frank Keller. Modeling human reading with neural attention. In *Proceedings of the 2016 Conference on Empirical Methods in Natural Language Processing*, pages 85–95, Austin, Texas, November 2016. Association for Computational Linguistics. https://doi.org/10.18653/v1/D16-1009. https://aclanthology.org/D16-1009.

17. James A Michaelov, Megan D Bardolph, Seana Coulson, and Benjamin K Bergen. Different kinds of cognitive plausibility: why are transformers better than rnns at predicting n400 amplitude? *arXiv preprint* arXiv:2107.09648, 2021.

18. Noortje J. Venhuizen, Matthew W. Crocker, and Harm Brouwer. Semantic entropy in language comprehension. *Entropy*, 21(12), 2019. ISSN 1099–4300. https://doi.org/10.3390/e21121159. https://www.mdpi.com/1099-4300/21/12/1159.

19. Geoffrey E Hinton and Ruslan R Salakhutdinov. Reducing the dimensionality of data with neural networks. *science*, 313(5786):504–507, 2006.

20. Tal Ridnik, Emanuel Ben-Baruch, Asaf Noy, and Lihi Zelnik-Manor. Imagenet-21k pretraining for the masses. In *Thirty-fifth Conference on Neural Information Processing Systems Datasets and Benchmarks Track (Round 1)*, 2021.

21. Matthew E. Peters, Mark Neumann, Mohit Iyyer, Matt Gardner, Christopher Clark, Kenton Lee, and Luke Zettlemoyer. Deep contextualized word representations. *CoRR*, abs/1802.05365, 2018. http://arxiv.org/abs/1802.05365.

22. Wilson L Taylor. "cloze procedure": A new tool for measuring readability. *Journalism quarterly*, 30(4):415–433, 1953.

23. Timo Schick and Hinrich Schütze. Exploiting cloze-questions for few-shot text classification and natural language inference. In *Proceedings of the 16th Conference of the European Chapter of the Association for Computational Linguistics: Main Volume*, pages 255–269, Online, April 2021. Association for Computational Linguistics. https://doi.org/10.18653/v1/2021.eacl-main.20. https://aclanthology.org/2021.eacl-main.20.

24. Pengfei Liu, Weizhe Yuan, Jinlan Fu, Zhengbao Jiang, Hiroaki Hayashi, and Graham Neubig. Pre-train, prompt, and predict: A systematic survey of prompting methods in natural language processing. *ACM Comput. Surv.*, 55(9), jan 2023. ISSN 0360-0300. https://doi.org/10.1145/3560815.

25. Aakanksha Chowdhery, Sharan Narang, Jacob Devlin, Maarten Bosma, Gaurav Mishra, Adam Roberts, Paul Barham, Hyung Won Chung, Charles Sutton, Sebastian Gehrmann, et al. Palm: Scaling language modeling with pathways. *arXiv preprint* arXiv:2204.02311, 2022.

26. Suchin Gururangan, Ana Marasović, Swabha Swayamdipta, Kyle Lo, Iz Beltagy, Doug Downey, and Noah A. Smith. Don't stop pretraining: Adapt language models to domains and tasks. In *Proceedings of the 58th Annual Meeting of the Association for Computational Linguistics*, pages 8342–8360, Online, July 2020. Association for Computational Linguistics. https://doi.org/10.18653/v1/2020.acl-main.740. https://aclanthology.org/2020.acl-main.740.

27. Brian Lester, Rami Al-Rfou, and Noah Constant. The power of scale for parameter-efficient prompt tuning. In *Proceedings of the 2021 Conference on Empirical Methods in Natural Language Processing*, pages 3045–3059, Online and Punta Cana, Dominican Republic, November 2021. Association for Computational Linguistics. https://doi.org/10.18653/v1/2021.emnlp-main.243. https://aclanthology.org/2021.emnlp-main.243.

28. James Kirkpatrick, Razvan Pascanu, Neil Rabinowitz, Joel Veness, Guillaume Desjardins, Andrei A Rusu, Kieran Milan, John Quan, Tiago Ramalho, Agnieszka Grabska-Barwinska, et al. Overcoming catastrophic forgetting in neural networks. *Proceedings of the national academy of sciences*, 114(13):3521–3526, 2017.

29. Neil Houlsby, Andrei Giurgiu, Stanislaw Jastrzebski, Bruna Morrone, Quentin De Laroussilhe, Andrea Gesmundo, Mona Attariyan, and Sylvain Gelly. Parameter-efficient transfer learning for nlp. In *International Conference on Machine Learning*, pages 2790–2799. PMLR, 2019.

30. Jonas Pfeiffer, Ivan Vulić, Iryna Gurevych, and Sebastian Ruder. MAD-X: An Adapter-Based Framework for Multi-Task Cross-Lingual Transfer. In *Proceedings of the 2020 Conference on Empirical Methods in Natural Language Processing (EMNLP)*, pages 7654–7673, Online, November 2020c. Association for Computational Linguistics. https://doi.org/10.18653/v1/2020.emnlp-main.617. https://aclanthology.org/2020.emnlp-main.617.

31. Jonas Pfeiffer, Andreas Rücklé, Clifton Poth, Aishwarya Kamath, Ivan Vulić, Sebastian Ruder, Kyunghyun Cho, and Iryna Gurevych. AdapterHub: A framework for adapting transformers. In *Proceedings of the 2020 Conference on Empirical Methods in Natural Language Processing: System Demonstrations*, pages 46–54, Online, October 2020b. Association for Computational Linguistics. https://doi.org/10.18653/v1/2020.emnlp-demos.7. https://aclanthology.org/2020.emnlp-demos.7.

32. Yinhan Liu, Myle Ott, Naman Goyal, Jingfei Du, Mandar Joshi, Danqi Chen, Omer Levy, Mike Lewis, Luke Zettlemoyer, and Veselin Stoyanov. Roberta: A robustly optimized BERT pretraining approach. *CoRR*, abs/1907.11692, 2019. http://arxiv.org/abs/1907.11692.

33. Alec Radford, Jeffrey Wu, Rewon Child, David Luan, Dario Amodei, Ilya Sutskever, et al. Language models are unsupervised multitask learners. *OpenAI blog*, 1(8):9, 2019.

34. Bernard Spolsky. Reduced redundancy as a language testing tool. In G.E. Perren and J.L.M. Trim, editors, *Applications of linguistics*, page 383-390. Cambridge University Press, Cambridge, 1969.

35. Zihang Dai, Zhilin Yang, Yiming Yang, Jaime Carbonell, Quoc Le, and Ruslan Salakhutdinov. Transformer-XL: Attentive language models beyond a fixed-length context. In *Proceedings of the 57th Annual Meeting of the Association for Computational Linguistics*, pages 2978–2988, Florence, Italy, July 2019. Association for Computational Linguistics. https://doi.org/10.18653/v1/P19-1285. https://aclanthology.org/P19-1285.

36. Zhilin Yang, Zihang Dai, Yiming Yang, Jaime Carbonell, Russ R Salakhutdinov, and Quoc V Le. Xlnet: Generalized autoregressive pretraining for language understanding. *Advances in neural information processing systems*, 32, 2019.

37. Frank Keller. Cognitively plausible models of human language processing. In *Proceedings of the 48th Annual Meeting of the Association for Computational Linguistics: Short Papers*, pages 60–67, 2010.

38. Keith Rayner and Arnold D Well. Effects of contextual constraint on eye movements in reading: A further examination. *Psychonomic Bulletin & Review*, 3(4):504–509, 1996.

39. Scott A McDonald and Richard C Shillcock. Eye movements reveal the on-line computation of lexical probabilities during reading. *Psychological science*, 14(6):648–652, 2003.

40. Hannah S Sarvasy, Adam Milton Morgan, Jenny Yu, Victor S Ferreira, and Shota Momma. Cross-clause planning in nungon (papua new guinea): Eye-tracking evidence. *Memory & Cognition*, pages 1–15, 2022.

41. Kevin Clark, Minh-Thang Luong, Quoc Le, and Christopher D. Manning. Pre-training transformers as energy-based cloze models. In *Proceedings of the 2020 Conference on Empirical Methods in Natural Language Processing (EMNLP)*, pages 285–294, Online, November 2020. Association for Computational Linguistics. https://doi.org/10.18653/v1/2020.emnlp-main.20. https://aclanthology.org/2020.emnlp-main.20.

42. Alexis Conneau, Douwe Kiela, Holger Schwenk, Loïc Barrault, and Antoine Bordes. Supervised learning of universal sentence representations from natural language inference data. In *Proceedings of the 2017 Conference on Empirical Methods in Natural Language Processing*, pages 670–680, Copenhagen, Denmark, September 2017. Association for Computational Linguistics. https://doi.org/10.18653/v1/D17-1070. https://aclanthology.org/D17-1070.

43. Emily M. Bender. Linguistically naïve != language independent: Why NLP needs linguistic typology. In *Proceedings of the EACL 2009 Workshop on the Interaction between Linguistics and Computational Linguistics: Virtuous, Vicious or Vacuous?*, pages 26–32, Athens, Greece, March 2009. Association for Computational Linguistics. https://aclanthology.org/W09-0106.

44. Marcell Fekete. Cross-lingual transfer using stacked language adapters. Master's thesis, Vrije Universiteit Amsterdam, 2022.

45. Rico Sennrich, Barry Haddow, and Alexandra Birch. Neural machine translation of rare words with subword units. In *Proceedings of the 54th Annual Meeting of the Association for Computational Linguistics (Volume 1: Long Papers)*, pages 1715–1725, Berlin, Germany, August 2016. Association for Computational Linguistics. https://doi.org/10.18653/v1/P16-1162. https://aclanthology.org/P16-1162.

46. Taku Kudo. Subword regularization: Improving neural network translation models with multiple subword candidates. In *Proceedings of the 56th Annual Meeting of the Association for Computational Linguistics (Volume 1: Long Papers)*, pages 66–75, Melbourne, Australia, July

2018. Association for Computational Linguistics. https://doi.org/10.18653/v1/P18-1007. https://aclanthology.org/P18-1007.

47. Linting Xue, Aditya Barua, Noah Constant, Rami Al-Rfou, Sharan Narang, Mihir Kale, Adam Roberts, and Colin Raffel. ByT5: Towards a token-free future with pre-trained byte-to-byte models. *Transactions of the Association for Computational Linguistics*, 10:291–306, 2022. https://doi.org/10.1162/tacl_a_00461. https://aclanthology.org/2022.tacl-1.17.

48. Yi Tay, Vinh Q Tran, Sebastian Ruder, Jai Gupta, Hyung Won Chung, Dara Bahri, Zhen Qin, Simon Baumgartner, Cong Yu, and Donald Metzler. Charformer: Fast character transformers via gradient-based subword tokenization. In *International Conference on Learning Representations*, 2021.

49. Sean Papay, Sebastian Padó, and Ngoc Thang Vu. Addressing low-resource scenarios with character-aware embeddings. In *Proceedings of the Second Workshop on Subword/Character LEvel Models*, pages 32–37, New Orleans, June 2018. Association for Computational Linguistics. https://doi.org/10.18653/v1/W18-1204. https://aclanthology.org/W18-1204.

50. Arij Riabi, Benoît Sagot, and Djamé Seddah. Can character-based language models improve downstream task performances in low-resource and noisy language scenarios? In *Proceedings of the Seventh Workshop on Noisy User-generated Text (W-NUT 2021)*, pages 423–436, Online, November 2021. Association for Computational Linguistics. https://doi.org/10.18653/v1/2021.wnut-1.47. https://aclanthology.org/2021.wnut-1.47.

51. Shotaro Misawa, Motoki Taniguchi, Yasuhide Miura, and Tomoko Ohkuma. Character-based Bidirectional LSTM-CRF with words and characters for Japanese Named Entity Recognition. In *Proceedings of the first workshop on subword and character level models in NLP*, pages 97–102, 2017.

52. Xinxiong Chen, Lei Xu, Zhiyuan Liu, Maosong Sun, and Huanbo Luan. Joint learning of character and word embeddings. In *Twenty-Fourth International Joint Conference on Artificial Intelligence*, 2015.

53. Wafia Adouane, Simon Dobnik, Jean-Philippe Bernardy, and Nasredine Semmar. A comparison of character neural language model and bootstrapping for language identification in multilingual noisy texts. In *Proceedings of the Second Workshop on Subword/Character LEvel Models*, pages 22–31, 2018.

54. SungMahn Ahn, Yeojin Chung, Jaejoon Lee, and Jiheon Yang. Korean sentence generation using phoneme-level LSTM language model. *Journal of Intelligence and Information Systems*, 23(2):71–88, 2017.

55. Sungjoon Park, Jeongmin Byun, Sion Baek, Yongseok Cho, and Alice Oh. Subword-level word vector representations for Korean. In *Proceedings of the 56th Annual Meeting of the Association for Computational Linguistics (Volume 1: Long Papers)*, pages 2429–2438, 2018.

56. Karl Stratos. A Sub-Character Architecture for Korean Language Processing. In *Proceedings of the 2017 Conference on Empirical Methods in Natural Language Processing*, pages 721–726, 2017.

57. Sangah Lee, Hansol Jang, Yunmee Baik, Suzi Park, and Hyopil Shin. Kr-BERT: A small-scale Korean-specific language model. *arXiv preprint* arXiv:2008.03979, 2020.

58. Phillip Rust, Jonas F Lotz, Emanuele Bugliarello, Elizabeth Salesky, Miryam de Lhoneux, and Desmond Elliott. Language modelling with pixels. *arXiv preprint* arXiv:2207.06991, 2022.

59. Richard Lawrence Lewis. *An architecturally-based theory of human sentence comprehension*. Carnegie Mellon University, 1993.

60. Peter Lindes and John E Laird. Toward integrating cognitive linguistics and cognitive language processing. In *Proceedings of the 14th International Conference on Cognitive Modeling (ICCM)*, 2016.

61. Shaojie Bai, J Zico Kolter, and Vladlen Koltun. An empirical evaluation of generic convolutional and recurrent networks for sequence modeling. *arXiv preprint* arXiv:1803.01271, 2018.

62. Sepp Hochreiter and Jürgen Schmidhuber. Long short-term memory. *Neural computation*, 9(8):1735–1780, 1997.

63. Ashish Vaswani, Noam Shazeer, Niki Parmar, Jakob Uszkoreit, Llion Jones, Aidan N Gomez, Ł ukasz Kaiser, and Illia Polosukhin. Attention is all you need. In I. Guyon, U. Von Luxburg, S. Bengio, H. Wallach, R. Fergus, S. Vishwanathan, and R. Garnett, editors, *Advances in Neural Information Processing Systems*, volume 30. Curran Associates, Inc., 2017. https://proceedings.neurips.cc/paper/2017/file/3f5ee243547dee91fbd053c1c4a845aa-Paper.pdf.

64. Will Bridewell and Paul Bello. A theory of attention for cognitive systems. *Advances in Cognitive Systems*, 4(1):1–16, 2016.

65. Moin Nadeem, Anna Bethke, and Siva Reddy. StereoSet: Measuring stereotypical bias in pre-trained language models. In *Proceedings of the 59th Annual Meeting of the Association for Computational Linguistics and the 11th International Joint Conference on Natural Language Processing (Volume 1: Long Papers)*, pages 5356–5371, Online, August 2021. Association for Computational Linguistics. https://doi.org/10.18653/v1/2021.acl-long.416. https://aclanthology.org/2021.acl-long.416.

66. Christine Basta, Marta R. Costa-jussà, and Noe Casas. Evaluating the underlying gender bias in contextualized word embeddings. In *Proceedings of the First Workshop on Gender Bias in Natural Language Processing*, pages 33–39, Florence, Italy, August 2019. Association for Computational Linguistics. https://doi.org/10.18653/v1/W19-3805. https://aclanthology.org/W19-3805.

67. Vijit Malik, Sunipa Dev, Akihiro Nishi, Nanyun Peng, and Kai-Wei Chang. Socially aware bias measurements for Hindi language representations. In *Proceedings of the 2022 Conference of the North American Chapter of the Association for Computational Linguistics: Human Language Technologies*, pages 1041–1052, Seattle, United States, July 2022. Association for Computational Linguistics. https://doi.org/10.18653/v1/2022.naacl-main.76. https://aclanthology.org/2022.naacl-main.76.

68. Pranav Narayanan Venkit, Mukund Srinath, and Shomir Wilson. A study of implicit bias in pre-trained language models against people with disabilities. In *Proceedings of the 29th International Conference on Computational Linguistics*, pages 1324–1332, Gyeongju, Republic of Korea, October 2022. International Committee on Computational Linguistics. https://aclanthology.org/2022.coling-1.113.

69. Maria Antoniak and David Mimno. Bad seeds: Evaluating lexical methods for bias measurement. In *Proceedings of the 59th Annual Meeting of the Association for Computational Linguistics and the 11th International Joint Conference on Natural Language Processing (Volume 1: Long Papers)*, pages 1889–1904, Online, August 2021. Association for Computational Linguistics. https://doi.org/10.18653/v1/2021.acl-long.148. https://aclanthology.org/2021.acl-long.148.

70. Hila Gonen and Yoav Goldberg. Lipstick on a pig: Debiasing methods cover up systematic gender biases in word embeddings but do not remove them. In *Proceedings of the 2019 Conference of the North American Chapter of the Association for Computational Linguistics: Human Language Technologies, Volume 1 (Long and Short Papers)*, pages 609–614, Minneapolis, Minnesota, June 2019. Association for Computational Linguistics. https://doi.org/10.18653/v1/N19-1061. https://aclanthology.org/N19-1061.

71. Allison Koenecke, Andrew Nam, Emily Lake, Joe Nudell, Minnie Quartey, Zion Mengesha, Connor Toups, John R Rickford, Dan Jurafsky, and Sharad Goel. Racial disparities in automated speech recognition. *Proceedings of the National Academy of Sciences*, 117(14):7684–7689, 2020.

72. Sarah Myers West, Meredith Whittaker, and Kate Crawford. Discriminating systems. *AI Now*, 2019.

73. Zeerak Waseem, Thomas Davidson, Dana Warmsley, and Ingmar Weber. Understanding abuse: A typology of abusive language detection subtasks. In *Proceedings of the First Workshop on Abusive Language Online*, pages 78–84, Vancouver, BC, Canada, August 2017. Association for Computational Linguistics. https://doi.org/10.18653/v1/W17-3012. https://aclanthology.org/W17-3012.

74. Luke Breitfeller, Emily Ahn, David Jurgens, and Yulia Tsvetkov. Finding microaggressions in the wild: A case for locating elusive phenomena in social media posts. In *Proceedings of the 2019 Conference on Empirical Methods in Natural Language Processing and the 9th International Joint Conference on Natural Language Processing (EMNLP-IJCNLP)*, pages 1664–1674, Hong Kong, China, November 2019. Association for Computational Linguistics. https://doi.org/10.18653/v1/D19-1176. https://aclanthology.org/D19-1176.

75. Emily M Bender, Timnit Gebru, Angelina McMillan-Major, and Shmargaret Shmitchell. On the dangers of stochastic parrots: Can language models be too big? In *Proceedings of the 2021 ACM Conference on Fairness, Accountability, and Transparency*, pages 610–623, 2021.

76. Alex Warstadt, Yian Zhang, Xiaocheng Li, Haokun Liu, and Samuel R. Bowman. Learning which features matter: RoBERTa acquires a preference for linguistic generalizations (eventually). In *Proceedings of the 2020 Conference on Empirical Methods in Natural Language Processing (EMNLP)*, pages 217–235, Online, November 2020b. Association for Computational Linguistics. https://doi.org/10.18653/v1/2020.emnlp-main.16. https://aclanthology.org/2020.emnlp-main.16.

77. Emma Strubell, Ananya Ganesh, and Andrew McCallum. Energy and policy considerations for deep learning in NLP. In *Proceedings of the 57th Annual Meeting of the Association for Computational Linguistics*, pages 3645–3650, Florence, Italy, July 2019. Association for Computational Linguistics. https://doi.org/10.18653/v1/P19-1355. https://aclanthology.org/P19-1355.

78. Roberta Michnick Golinkoff, Erika Hoff, Meredith L Rowe, Catherine S Tamis-LeMonda, and Kathy Hirsh-Pasek. Language matters: Denying the existence of the 30-million-word gap has serious consequences. *Child development*, 90(3):985–992, 2019.

79. Douglas E Sperry, Linda L Sperry, and Peggy J Miller. Reexamining the verbal environments of children from different socioeconomic backgrounds. *Child development*, 90(4):1303–1318, 2019.

80. Manja Attig and Sabine Weinert. What impacts early language skills? effects of social disparities and different process characteristics of the home learning environment in the first 2 years. *Frontiers in Psychology*, 11:557751, 2020.

81. Betty Hart and Todd R Risley. The early catastrophe. *Education review*, 17(1), 2003.

82. Philip A. Huebner, Elior Sulem, Fisher Cynthia, and Dan Roth. BabyBERTa: Learning more grammar with small-scale child-directed language. In *Proceedings of the 25th Conference on Computational Natural Language Learning*, pages 624–646, Online, November 2021. Association for Computational Linguistics. https://doi.org/10.18653/v1/2021.conll-1.49. https://aclanthology.org/2021.conll-1.49.

83. Yian Zhang, Alex Warstadt, Xiaocheng Li, and Samuel R. Bowman. When do you need billions of words of pretraining data? In *Proceedings of the 59th Annual Meeting of the Association for Computational Linguistics and the 11th International Joint Conference on Natural Language Processing (Volume 1: Long Papers)*, pages 1112–1125, Online, August 2021. Association for Computational Linguistics. https://doi.org/10.18653/v1/2021.acl-long.90. https://aclanthology.org/2021.acl-long.90.

84. Jochen L. Leidner and Vassilis Plachouras. Ethical by design: Ethics best practices for natural language processing. In *Proceedings of the First ACL Workshop on Ethics in Natural Language Processing*, pages 30–40, Valencia, Spain, April 2017. Association for Computational Linguistics. https://doi.org/10.18653/v1/W17-1604. https://aclanthology.org/W17-1604.

85. European Commission. Independent high-level expert group on artificial intelligence: Ethics guidelines for trustworthy ai. *Search in*, 2019. https://www.aepd.es/sites/default/files/2019-12/ai-ethics-guidelines.pdf.

86. Anna Rogers. Changing the world by changing the data. In *Proceedings of the 59th Annual Meeting of the Association for Computational Linguistics and the 11th International Joint Conference on Natural Language Processing (Volume 1: Long Papers)*, pages 2182–2194, Online, August 2021. Association for Computational Linguistics. https://doi.org/10.18653/v1/2021.acl-long.170. https://aclanthology.org/2021.acl-long.170.

Cognitive Signals of Language Processing

Language comprehension is an automated task that we perform all the time and usually effortlessly. If we hear people talk in our native language or see a written sign, it would be almost impossible for us not to listen to them or not to read and interpret the sign. When we study language processing, we want to understand how the elementary components of language are combined to derive meaning during such comprehension acts.

Cognitive signals of language processing can help us to compare computational models to human processing. The core idea behind using cognitive processing signals in natural language processing is to capture measurable observations of subconscious phenomena of language understanding. Cognitive signals come in many forms: from behavioral data such as annotation rationales and response times to online processing metrics such as eye-tracking data and complex brain scans. They can be captured during auditory or visual language processing and can be analyzed as measurable reflections of the cognitive functions involved in the process. A better understanding of human language processing will help us develop cognitively more plausible natural language processing models. Conversely, computational modeling forces us to operationalize psycholinguistically underspecified concepts explicitly and the experimental results can lead to theoretical adjustments.

The main challenges in working with cognitive signals of language processing are the limited size and availability of datasets and the inherent noisiness of the data. When we use cognitive signals to inspect the cognitive plausibility of language models, we need to find methodological solutions to separate signals pertaining to language processing from interfering noise caused by other cognitive processes happening at the same time [1]. For instance, the fMRI signal contains neural activation responses for thousands of voxels distributed over the whole brain mapped to a three-dimensional spatial coordinate system. The signal needs to undergo multiple statistical preprocessing filters in an attempt to separate the

L. Beinborn and N. Hollenstein, *Cognitive Plausibility in Natural Language Processing*, Synthesis Lectures on Human Language Technologies, https://doi.org/10.1007/978-3-031-43260-6_3

patterns related to language processing from patterns caused by other cognitive processes that are happening simultaneously such as distracting thoughts and muscle movements [2].

In this chapter, we introduce various types of cognitive signals and describe how they reflect human language processing (Sect. 3.1). We further discuss necessary preprocessing steps and the challenges that arise when working with cognitive signals (Sect. 3.2). To exemplify the advantages and disadvantages of cognitive datasets, we zoom in on some representative examples and outline how they can be used in NLP research (Sect. 3.3). Cognitive signals can be used to enrich the representations or the inductive bias of language models or to evaluate their cognitive plausibility through a range of interpretability methods. We will expand on the experimental options in more detail in the following chapters. As cognitive signals are extracted from human participants, collecting and processing them requires a high degree of ethical sensitivity (Sect. 3.4).

3.1 Cognitive Signal Types

Cognitive processing signals can be distinguished into online and offline metrics. We focus on cognitive signals recorded during language comprehension rather than language production, and on signals recorded from adult participants. We introduce different types of cognitive data and describe how they can be used to test and improve the cognitive plausibility of natural language processing models along the three analysis dimensions: behavioral patterns, representational structure, and processing strategies.

Behavioral patterns are usually studied by comparing them to offline signals such as annotations, ratings, or answers, that have been explicitly and consciously produced by humans upon request and provide a cumulative impression of the cognitive load associated with the task. Representational structure and processing strategies can better be examined using online signals, which are recorded from humans in real-time and are effortlessly produced as a side product of language processing. Online signals include eye-tracking fixations, keystroke and scrolling logs, and brain activity data.

In Table 3.1, we collected introductory references that provide more details about the respective signal types.

3.1.1 Offline Measures

Offline measures are data collected from explicit human judgments. The most established means for approximating human behavior in natural language processing are annotations. Annotations are commonly stored as metadata providing additional information about an instance. Humans are usually seen as the ultimate gold standard for language processing and NLP models are evaluated by how well they can reproduce human annotations on specific tasks. Wilcock [10] provides a comprehensive introduction to linguistic annotation. The

Table 3.1 Introductory references for the cognitive signal types

Signal	Introductory references
Annotation	Pustejovsky and Stubbs [3]
Self-paced reading	Jegerski [4]
Eye-tracking	Holmqvist et al. [5]
EEG	Cohen [6]
MEG	Hansen et al. [7]
fMRI	Soares et al. [8]
fNIRS	Quaresima et al. [9]

main advantage of most offline methods is that they can be executed over the web with participants recruited from crowd-sourcing platforms. This procedure leads to fast and low-cost data collection and allows for larger datasets and a more representative participant population.

From a psycholinguistic perspective, we are additionally interested in the difficulty that is associated with processing or annotating an instance. The difficulty of an instance can be approximated by calculating annotator agreement or by measuring the cognitive load it causes. From a behavioral perspective, cognitive load can be approximated by accumulated reading and response times [11].

3.1.1.1 Annotator Agreement

Human annotation decisions are prone to inconsistencies due to under-specified annotation guidelines, ambiguity in the samples, insufficient annotator expertise, human errors, and subjectivity, among other reasons [12, 13]. The inter-annotator agreement over all instances is a means to quantify such inconsistencies [14]. In natural language processing, disagreement between annotators is traditionally interpreted as a complication in the annotation process, and different resolution strategies are applied to resolve conflicts [15, 16]. The most common strategy is majority voting which assigns the label that has been chosen by the majority of the annotators.

From a complementary perspective, the overall inter-annotator agreement can be interpreted as an explicit indicator of the difficulty of a task that determines the human upper bound [17]. Accordingly, the agreement for a single instance can be interpreted as the difficulty of the instance (we discuss disagreement and difficulty in more detail in Sect. 4.1.2.2). This interpretation assumes that all annotators have equal expertise on the task and ignores annotator biases. In contrast, Rasch models of difficulty are based on the item-response theory and consider the ability of the annotator as a latent trait [18]. Rasch models are supervised and estimate the difficulty by using the correct label for an instance which is usually not known for annotation tasks. Alternatively, Bayesian models of annotation are able to estimate the difficulty in an unsupervised fashion [19].

3.1.1.2 Human Annotation Rationales

Human annotators might assign the same annotation label to an instance despite interpreting it differently. In order to compare the reasoning processes that led to an annotation, human rationales can be collected by asking annotators to highlight the elements of the input that are relevant to their output decision [20]. Rationales reveal important information about human explanations and are often compared with local attribution signals in language models [21]. Søgaard [22] distinguishes between extractive rationales (highlighted regions of instances) and abstractive rationales (free text explanations). The latter are comparable to think-aloud protocols [23]. Human rationales provide conscious post-hoc explanations, which might not properly reflect the subconscious processing patterns and need to be interpreted cautiously [24].

3.1.1.3 Human Response Times

Recording human response times is a simple and inexpensive method to approximate cognitive load without requiring specialized equipment such as eye trackers or neuroimaging devices. Longer response times can be interpreted either as a sign of uncertainty of the participant or as higher difficulty of the instance. Response times are particularly interesting for providing additional information on subjective tasks for which human judgments cannot be deemed correct or false such as word similarity or analogy ratings. They are a popular means to investigate semantic priming effects [25, 26]. In priming experiments, participants are usually asked to perform a lexical decision task which means that they decide whether a string of characters forms an existing word. For example, the sequence *grill* should be identified as a real English word and *grell* as a pseudo-word. Response times for this task vary when participants are presented with a semantically related word—a prime—before taking the decision. These differences in response times are attributed to characteristics of the prime that facilitate or complicate lexical access. For example, the lexical decision for *grill* is expected to be faster, when *heat* or *barbecue* were presented as primes. We further discuss priming phenomena in Sect. 6.2.2.2.

Incremental variants of the cloze task [27] have been developed to analyze response times for different conditions of incremental sentence processing. In cloze tasks, the predictability of a word in a context is determined by presenting participants with the beginning of a sentence or a passage p to the word of interest and asking them to produce the word that would be most likely to follow [28, 29]. This approach is similar to the next-word prediction objective for language modeling. The proportional responses provided by the participants can be viewed as annotator agreement for the cloze task and as an indicator of context specificity. The maze task is a more controlled variant of this incremental procedure [30]. Participants read a sentence word by word and are presented with a forced choice at each position. They are asked to distinguish between a correct word that leads to a reasonable continuation of the sentence and a contextually inappropriate distractor. As the selection of distractors is a laborious task, Beinborn et al. [31] and Boyce et al. [32] discuss how the suitability of distractors can be automatically assessed using NLP methods. While participants in cloze and maze tasks are asked to produce explicit, conscious judgments, they are instructed to respond as quickly as possible because their response times are tracked as dependent measures of cognitive load. These fast-paced incremental decisions can also give insights into procedural phenomena indicating that the distinction between offline and online measures should rather be understood as a continual range. While annotations can only provide information about the final outcome of a conscious decision, response times can indicate cognitive load but it remains unclear what the participants were doing. They might simply have taken a break from the task. More direct online signals can provide localized information about the trigger that causes longer response times and can reveal subconscious cognitive patterns such as biases or naive heuristics that lead to simplified decision processes.

3.1.2 Online Measures

Online methods such as keystroke metrics, scrolling logs, and fixation patterns recorded by eye-tracking devices capture procedural signals in real-time. They can provide insights into processing patterns such as attention, back-tracking paths, and the relative importance of the input tokens.

3.1.2.1 Cursor and Keystroke Metrics

Cursor movements and scrolling logs provide important information about the readability of a text and reveal significant differences in the reader's interactions with the text [33]. These signals are semi-consciously produced and can be collected inexpensively without interfering with the reading process.

Keystroke metrics are predominantly used to study written language production. The latency between successive keystrokes captures the user's typing patterns and can indicate cognitive load and repair processes. In this book, we focus mainly on language comprehension but cognitive processing signals produced during writing can also serve as a source of information for syntactic tasks. Plank [34] show that keystroke dynamics carry information about structural relations between words that can be used to improve syntactic parsing. Kerz et al. [35] study second language writing and link keystroke logging data to indices of syntactic and lexical complexity of the produced texts. Their methodology establishes a promising link between process and product analyses of literacy.

3.1.2.2 Self-paced Reading

Self-paced reading is an experimental paradigm that enforces sequential word-by-word (or phrase-by-phrase) reading. Participants need to actively press a button or perform a mouse click to trigger the display of the next word. In this moving window approach, the current word is hidden when revealing the next one which makes it possible to record the time the reader required for each word (i.e., the time a word was visible). This time span gives an indication of the processing difficulty experienced at the respective word and is often interpreted as cognitive load [4, 11]. A variation is the cumulative window paradigm in which the previously presented words remain visible and can be re-read.

The main advantage of self-paced reading is that the experiments can easily be scaled to crowd-sourcing settings with large numbers of participants as no specialized equipment is required. Self-paced reading data can be noisy compared to eye-tracking signals since backtracking is not allowed and the attention of the participant cannot be assessed (i.e., we cannot control whether the participant is looking at the word when it is presented), but it has been an important source of information of incremental processing difficulty [36, 37].

3.1.2.3 Eye-Tracking

When we read, our eyes move rapidly over the text in sequences of fixations. Some words are not fixated at all due to an intricate interplay of preview and predictability effects, and some words are fixated several times due to factors such as syntactic re-analysis [38]. Eye-tracking signals are recorded with a device that tracks the movements of the pupil in a non-intrusive way. The most accurate devices emit infrared light and record the light reflections on the pupil with a camera. A gaze estimation algorithm then extracts gaze patterns by smoothing over the captured movements. Devices with a high sampling rate (200 Hz or more) can provide very fine-grained temporal records of the gaze patterns of one or both eyes indicating different cognitive stages of reading.

Early gaze measures such as first fixation duration and first pass duration are measured when a word is fixated for the first time and capture lexical access and early syntactic processing. Around 10–15% of the fixations are regressions which means that the eye focus jumps back to re-read a part of the text. Late measures are extracted when words are fixated more than once and reflect processes of disambiguation and syntactic re-analysis. A fixation lasts on average around 200 ms with large standard deviations. The duration of a fixation is influenced by the semantic, syntactic, and discourse-related properties of the text being read.

Eye-tracking technology is becoming cheaper and more easily available by the day [39, 40]. Currently, gaze patterns are still mostly recorded in controlled experiment environments but even the cameras of mobile devices can now record gaze data with relatively robust quality [41, 42]. Guan et al. [43] use webcam-based eye-tracking to examine the reading comprehension processes of English L2 speakers. Their results show that the number of fixations on specific textual elements was predictive of the participant's answer to text comprehension questions. Methodological advances in webcam-based eye-tracking can combine the benefits of high-quality eye-tracking with the scalability of self-paced reading experiments and facilitate the creation of sizable eye-tracking datasets [44]. For remote experimental scenarios, additional measures have to be taken to assure the transparent communication of the experimental conditions and the privacy of the participants.

Eye-tracking has been used as a proxy for cognitive load and its integration in NLP models has led to improvements for a wide range of tasks [45]. Word-level eye-tracking features provide useful theory-independent measures to investigate sentence comprehension (e.g., Demberg and Keller [46]) and recently have been shown to be predictive of neural activity in the language network [47]. While the quantitative gains in performance remain modest, they have been reported for multiple tasks including part-of-speech tagging [48], dependency parsing [49], sentiment analysis [50–52], and text summarization [53]. We discuss methods to evaluate and improve the cognitive plausibility of language models using eye-tracking data in more detail in Sect. 6.3. Due to the many open pre-processing challenges in this relatively new area that we describe in Sect. 3.2, it remains to be seen whether the observed tendencies can be consistently reproduced across a variety of models and datasets.

3.1.3 Brain Activity Data

The signal type that comes closest to measuring subconscious cognitive processing is brain activity data. Collecting it requires the most expensive experimental equipment, rigid experimental conditions, and a combination of several statistical preprocessing tools. Being able to explore brain activity data awakens inspiring hopes but properly interpreting the complex signal is a tedious and sometimes frustrating task. The most established methods for measuring brain activity are EEG, MEG, fMRI, and fNIRS.

3.1.3.1 M/EEG
The electrical activity of neurons produces currents spreading through the brain. These currents also reach the scalp and the resulting voltage fluctuations can be captured as electroencephalography recordings (EEG) from the surface. For this, a cap with a set of electrodes at different positions around the head is placed on the scalp of the participants. Each electrode will capture the brain activity reaching its location. EEG signals reflect electrical brain activity with a millisecond-accurate temporal resolution, but poor spatial resolution. The high temporal resolution of EEG allows for more fine-grained language understanding experiments on the word level, which is crucial for natural language processing [54]. EEG signals can be split into frequency bands which can be linked to certain linguistic observations. For instance, effects related to semantic violations can be found within the gamma frequency range ($\sim$30–100 Hz), with well-formed sentences showing higher gamma levels than sentences containing violations [55].

The neuronal currents produce magnetic fields that can be measured on the scalp surface and captured as a magnetoencephalogram (MEG). Like EEG, the MEG signal tracks neural activity as it happens, millisecond by millisecond, but it also offers higher spatial resolution as magnetic fields can be used to triangulate the electrical source of activity in the brain.

Magnetic fields are less distorted than electric fields by the skull and scalp which results in a better spatial resolution for MEG.

M/EEG data contains many types of artifacts caused by interferences from other cognitive processes. Artifacts have different sources such as eye movements and blinks, or muscular movements from other body parts including the heartbeat. These artifacts can be removed automatically [56, 57], and EEG signals that have been time-aligned to an event such as the onset of a stimulus presentation can be averaged as event-related potentials (ERPs).

In language processing, the most established EEG phenomenon is the N400 component [58]. The N400 is a negative deflection in the event-related brain potential which peaks approximately 400 ms after the presentation of a stimulus [59]. The N400 is part of the common brain response to words and other meaningful stimuli, including visual and auditory linguistic input. It is associated with processing difficulty: If the preceding context activates semantic associations related to an upcoming word, the word is easier to process and thus elicits a reduced N400. Psycholinguistic studies indicate that the N400 component specifically indexes the extent to which the upcoming word was not expected; i.e., the surprisal [60–62].

The cognitive plausibility of language models can be evaluated by their ability to predict EEG effects linked to syntactic and semantic phenomena [63–66]. We review such experiments when discussing the cognitive plausibility of procedural strategies in Chap. 6. The main advantages of using M/EEG for analyzing language processing are the temporal resolution and the lower number of dimensions compared to fMRI signals. Due to the low signal-to-noise ratio in the EEG data, machine learning techniques have been applied to automatically preprocess the signal [67–69].

3.1.3.2 FMRI and FNIRS

Functional magnetic resonance imaging (fMRI) and functional near-infrared spectroscopy (fNIRS) are both neuroimaging techniques that estimate hemodynamic activity in the brain. FMRI is a neuroimaging technique that measures brain activity by the changes in the oxygen level of the blood. This technique relies on the fact that cerebral blood flow and neuronal activation are coupled: When a brain area is in use, blood flow to that area increases. In other words, when neurons in a given brain region become active, oxygen-rich blood replaces oxygen-depleted blood a few seconds later.

FMRI produces 3D scans of the brain with a high spatial resolution of the signal. For statistical analyses, the brain scan is fragmented into voxels which are cubes of constant size. The signal is interpreted as an activation value for every voxel. The number of voxels varies depending on the precision of the scanner and the size and shape of the participant's brain. The voxel location can be identified with 3-dimensional coordinates, but the signal is commonly processed as a flattened vector which ignores the spatial relationships between the voxels. This rather naive modeling assumption simplifies the signal but might lead to cognitively and biologically implausible findings.

Most publicly available fMRI datasets have already undergone common statistical filters. These preprocessing steps correct for the participant's head movements, account for different timing of the scan slices, and adjust linear trends in the signal [70]. In addition, the scans of the individual brains (which vary in size and shape) need to be aligned with a standardized template to group voxels into brain regions and allow for comparisons across participants. Researchers using datasets that have been collected and published by another lab should be aware of the effect of these probabilistic corrections. They are necessary to further analyze the signal, but might also systematically add noise to the data and lead to misinterpretations.

The fMRI signal measures a brain response to a stimulus with a delay of a few seconds, which decays slowly over a duration of several seconds [71]. When processing fMRI data, this hemodynamic delay should be taken into account. For continuous language stimuli, this means that the response to previous stimuli will have an influence on the current signal.

Although working with brain imaging data is highly promising from a cognitive perspective, it comes with many practical limitations. Brain datasets are commonly too small to train powerful machine learning models, the imaging technology produces noisy output that needs to be adjusted by statistical correction methods, and most importantly, only very few datasets are publicly available (see Sect. 3.3 for some examples).

Recordings of brain activity play a central role in furthering our understanding of how human language works and how concepts are represented in the brain [72]. In Sect. 5.2.2, we discuss experiments to learn mapping functions between brain activation responses and computational representations of stimuli [73]. Such language-brain encoding experiments evaluate the ability of language models to predict brain responses elicited by language stimuli and are a promising direction for evaluating the cognitive plausibility of models. However, the evaluation scenarios for this task have not yet been standardized which makes it difficult to compare and interpret results [74].

The signal captured by functional near-infrared spectroscopy (fNIRS) is similar to fMRI responses as it also captures changes in blood oxygen concentration. For this method, brain activity is measured by using near-infrared light to estimate cortical hemodynamic activity that occurs in response to neural activity. FNIRS is a non-invasive technique that can even be used in portable contexts but it only captures signals from brain regions near the cortical surface. Quaresima et al. [9] provide an overview of the strengths and weaknesses attributed to the fNIRS technique.

FNIRS signals can be used to measure the connectivity of language-related areas in the brain, for example, to study the effect of cochlear implants [75]. Scherer et al. [76] show that they provide useful information for studying syntactic processing in the bilingual brain. Cao et al. [77] publish an fNIRS dataset and provide results for a brain decoding task. They record fNIRS signals on the participant's scalp while presenting images of objects as stimuli and train a decoder to take brain scans as input and output the corresponding text semantic vectors. The authors find that activation patterns recorded from fNIRS encode rich information for discriminating concepts, but show limitations on the possibility of decoding fine-grained semantic clues when compared to fMRI signals. Zhao et al. [78] use carefully designed stimuli to examine the relationship between syntactic analysis and semantic integration in Japanese reading using fNIRS signals. They compare the differences in brain responses for sentences in which the verb has been substituted with a semantically unrelated verb.

An important aspect to take into account when choosing an adequate cognitive signal type is the linguistic level of the analysis. Due to the low temporal resolution and hemodynamic delay of fMRI, eye-tracking or M/EEG signals are more useful for extracting word-level signals in continuous stimuli. When comparing signals from multiple datasets, it is crucial to ensure that the statistical correction methods that have been applied to the datasets are comparable.

3.1.4 Combining Signal Types

Any cognitive dataset only allows us to peek into a tiny aspect of human language processing of a small sample of participants. Cognitive signals of language processing are subject to individual differences and might therefore inhibit sample-specific biases. In order to draw more robust conclusions about cognitive processing phenomena, plausibility experiments can be compared across multiple datasets of the same cognitive modality to factor out the influence of the choice of stimuli and participant sample. For instance, Hollenstein and Zhang [79] combine gaze features from three corpora, and Mensch et al. [80] learn a shared representation across several fMRI datasets.

Cognitive signals are often intermingled with cues that are not directly relevant to language processing (e.g., motor control, vision). The sensitivity to noise varies across the different signal types. As different recording methods are complementary with respect to the temporal and spatial resolution of the signal, the combination of methods can increase the robustness of cognitive analyses. For instance, Schwartz et al. [81] used both MEG and fMRI data to enrich language representations and show how the integration of both signal types simultaneously improves their predictions. Hollenstein et al. [82] presented a unified framework for evaluating the cognitive plausibility of word embeddings by predicting eye-tracking, EEG, and fMRI signals from 15 different datasets. Their results show clear correlations between these three signal types. Barrett et al. [83] combined eye-tracking features with prosodic features, keystroke logs from different corpora, and pre-trained word embeddings for part-of-speech induction and chunking.

For some corpora, data was recorded from multiple modalities at different times on different participants, but on the same stimulus: For example, the UCL corpus [37] contains self-paced reading times and eye-tracking data, and was later extended with EEG data [84]. Similarly, self-paced reading times and fMRI were recorded for the Natural Stories Corpus [85, 86]; EEG and fMRI were recorded for the Alice corpus [64, 87].

For some sources, data from co-registration studies is available which means that two types of signals are recorded simultaneously during the same experiment. This has become more popular since EEG, fMRI and eye-tracking are complementary in terms of temporal and spatial resolution as well as the directness in the measurement of neural activity [88]. Co-registration studies for reading have been conducted to combine eye-tracking information with EEG [89–91] and fMRI signals [92]. These co-registration studies make it possible to compare the characteristics of different recording methods on the same language stimuli, the same population, and the same language understanding task. The presented recording modalities of cognitive signals in this paper are complementary to each other, the information provided by each modality adds to the full picture.

EEG is a recording technique with very high temporal resolution (i.e., resulting in multiple samples per second). However, as the electrodes measure electrical activity at the surface of the brain—through the bone—it is difficult to know exactly in which brain region the signal originated. EEG signals have been used frequently for classification in brain-computer-interfaces, e.g., classifying text difficulty for speech recognition [93], but have rarely been used to improve NLP tasks such as sentiment analysis [94, 95]. As there is only limited EEG data from naturalistic reading openly available, many processing decisions regarding the selection of features remain an open research question to date. The MEG technology yields better temporal and spatial resolution, which makes it very attractive for NLP but even fewer datasets from naturalistic studies are currently available. Finally, the fMRI signal exhibits opposite characteristics. Due to the precise 3D scans, the spatial resolution is very high; but, since it takes a few seconds to produce a scan over the full brain, the temporal resolution is very low. Recently, fMRI data has become popular in NLP to evaluate neural language models (e.g., Schwartz et al. [81]) and to improve word representations [96]. It is useful to leverage fMRI signals if the localization of cognitive processes plays an important role and to investigate theories about specialized processing areas. Unfortunately, fMRI scans are less accessible and more expensive than other signal types.

As cognitive signals contain considerable portions of noise, it is advisable to work with multiple datasets of the same modality or to compare multiple modalities to arrive at more robust interpretations. When using brain activity data, we always run the risk of reverse inference. This refers to the tendency to falsely infer details about cognitive processes from patterns of activation revealed by recording techniques. As any brain region can be involved in multiple processes at the same time, causal links between observed brain activity and task requirements are hard to establish [97].

When evaluating models augmented with cognitive signals, cross-dataset or cross-signal evaluation setups yield more reliable indications of cognitive plausibility. For example, Hollenstein and Zhang [79] train named entity recognition models on one eye-tracking corpus and test on another one. Schwartz et al. [81] compare the cognitive plausibility of language models for both fMRI and MEG data. Within the same type of cognitive signals, it is useful to compare models on data coming from differing experimental conditions (human language processing of spoken vs. written stimuli, different recording devices, different preprocessing steps, etc.).

3.2   Preprocessing Cognitive Signals for NLP

Human language processing data comes as a highly heterogeneous mix of linguistic, behavioral, and physiological datasets of limited size. Hence, extracting the linguistic structure and eliminating noise from other parallel cognitive processes is the first challenge in cognitively-inspired NLP [1, 45]. We describe different techniques for preprocessing cognitive data (e.g., noise correction, aggregation over participants, dimensionality reduction, alignment with stimuli) and their effect on NLP analyses.

3.2.1   Participant Aggregation

When working with cognitive processing signals, the recorded data from human individuals will differ and we need to find an efficient approach to combine data from multiple participants. A simple but naive starting point lies in aggregating an average signal across all participants. When using eye-tracking patterns and EEG signals to adjust the inductive bias of task-specific models, averaging can lead to more robust results [98]. It should be noted though that averaged signals often result in cognitively implausible patterns that do not correspond to any individual profile.

Alternatively, models can be trained on the data of individual participants. If the goal is to build generalizable models for application scenarios that can successfully deal with data from previously unseen participants, cross-participant evaluation is recommendable [99]. The variability between participants is often underestimated when working with averaged signals [100]. It has been shown that noisy data, for instance from participants with increased head movements, leads to significantly lower prediction accuracy in brain decoding tasks [73]. Furthermore, it is now also standard practice in psychology not to average blindly across participants, but to use techniques that take such random factors into account during data analysis, for instance, linear mixed-effects models (e.g., Schielzeth et al. [101]).

Cross-participant or cross-population evaluations are informative complementary analyses to the common approaches of taking the mean across participants or training individual-participant models. While high accuracy can be achieved on within-participant models, the performance usually drops for cross-participant evaluations. Brandl and Hollenstein [102] evaluate the impact of subgroup characteristics reflected in reading on the evaluation of the cognitive plausibility of models and find significant differences. When training machine learning models on cognitive processing signals, the patterns learned for a given task should ideally hold for the entire population. Leave-one-participant-out evaluation scenarios can provide important information on inter-participant variability [103]. With increasingly available datasets, cross-population experiments will be facilitated to compare, for example, language processing in native and non-native speakers.

3.2.2 Stimulus Alignment

The alignment between experimental stimuli and tokenized text in a format that is usable as the input for neural language models is one of the first challenges in any NLP project involving cognitive processing signals. However, it is rarely addressed and best practices are lacking [45]. It starts with the simple question of whether to apply standardized casing to the stimuli. While this is simply an experimental choice, the alignment of computational subtokens to experimental stimuli and the integration of response delays require more complex procedures and their impact on the results is currently understudied.

3.2.2.1 Tokenization in Eye-Tracking Data

Subword tokenizers handle unknown tokens by recursively splitting every word until all subtokens belong to its vocabulary. For example, the adjective *sensational* is tokenized into the three subtokens *['sen', '##sation', '##al']*. When working with eye-tracking data, one needs to decide how to map word-level fixation features to subtokens. In earlier work, we have experimented with different variants that both underestimate morphological effects: we used only the first subtoken when predicting eye-tracking signals [104], or we summed over attributions for subtokens to determine relative importance [105].

In eye-tracking studies, punctuation is usually not separated from the preceding token (*Sensational!*) and it remains an open question how to split the eye-tracking signal over the two components (*Sensational* and *!*). Some approaches assign zero or varying default values to the second token [79] or duplicate the eye-tracking values [106]. Both approaches skew the data and the impact of these preprocessing decisions has not yet been studied. With larger amounts of data, one could learn the alignment between the cognitive signals and continuous textual stimuli directly.

The alignment between gaze patterns and input tokens is further complicated by spill-over effects. As our response to language is not instantaneous [107, 108], the processing of adjacent words overlaps, and the target region for observing a certain effect may lag behind the stimulus that triggers it. In an English reading study, Smith and Levy [109] find that spill-over effects can impact a temporal span of up to three words after the target. While disregarding spill-over effects still permits the discovery of the relevant patterns for relatively static phenomena [110, 111], more caution is required when working with contextualized models.

3.2.2.2 Delay in FMRI Data

The hemodynamic delay refers to the time it takes between the onset of a stimulus and a measurable increase in the oxygen level of the blood in a certain brain region that can be picked up by an fMRI scanner. The duration of this delay varies across participants. As the hemodynamic response decays slowly over a duration of several seconds, the problem of overlapping responses for continuous stimuli is even more prevalent for fMRI data.

A naive hack for this problem consists in aligning the scans to the stimuli with a fixed offset of four seconds [73, 74]. Wehbe et al. [112] use a feature-based representation and learn different weights for stimuli occurring at previous time steps. The delay can also be approximated with linear filters. For example, Huth et al. [72] and Jain et al. [113] concatenate multiple channels corresponding to a delay of two, four, six, and eight seconds using linear weights. One channel represents the words processed two seconds earlier, another channel maps to a delay of four seconds earlier, and so on. In addition to the delay in the brain response, we also need to account for the delay caused by the scanner as each fMRI scan takes two seconds during which the participant is processing multiple words. The scanner delay is the amount of time between displaying the first word and completing the fMRI scan [114].

3.2.3   Dimensionality Reduction

Brain activity data is represented as a high-dimensional matrix and contains significant portions of noise from brain regions that are not related to the linguistic processes, we want to analyze. In order to extract a more robust representation of the signal, techniques for dimensionality reduction are often applied when using cognitive signals in natural language processing. We shortly introduce principal component analysis for feature extraction from EEG data and voxel selection in fMRI as examples of dimensionality reduction methods.

3.2.3.1 Component Analysis for EEG Data

Principal Component Analysis (PCA) and Independent Component Analysis (ICA) as well as a combination of these methods can be applied to the task of eliminating artifacts from the EEG signal.

PCA gathers the most possible channel activity into each component. Usually, this approach sums activity in multiple independent sources, leaving the remaining variance to be accounted for by subsequent principal components. Therefore, PCA is efficient for squashing the maximum variance in the data (regardless of the source) into a lower number of dimensions of features [115].

Independent Component Analysis, on the contrary, splits the channel activity into as many independent components as possible. ICA-based artifact correction can separate and remove a wide variety of artifacts from EEG data by linear decomposition. The ICA method is based on the assumption that the time series recorded on the scalp are spatially stable mixtures of the activities of temporally independent cerebral and artifactual sources. In this model, independence signifies that the information carried by one component cannot be inferred from the others [116].

ICA has been found to be an effective data-driven method for separating signals from temporally and functionally independent brain and non-brain source processes. Even if the sources are not exactly independent, ICA-based algorithms have been reported to be successful at removing different types of artifacts from the EEG signal of interest [117].

Independent components can be extracted from the mixed signals by using this method, even in free-viewing experiment paradigms [118].

3.2.3.2 FMRI Voxel Selection

Recall that fMRI scans represent neural activity as separate signals in millimeter-sized cubes called voxels. The number of voxels in a brain varies depending on the specified voxel size and the shape of the participant's brain. FMRI studies typically measure responses across the whole brain, but not all regions are likely to be related to language processing. Changes in neural activity might be independent of the stimuli but be affected by physical processes like the noise perception in the scanner. It is therefore common practice to apply voxel selection methods to analyze only a relevant subset of voxels.

In NLP applications, Hollenstein et al. [82] showed that when selecting voxels randomly, word embeddings perform better at predicting neural activation on a larger number of voxels. Analogously, Beinborn et al. [74] confirmed that random selection does not lead to improvements by simply reducing the dimensionality.

Theory-driven methods restrict the number of voxels by using knowledge from previous neuroscience research about brain regions of interest (e.g., Brennan et al. [87]). In contrast, information-driven voxel selection methods reduce the number of voxels by analyzing the predictive power of the voxel signal within a given subset and for a specific task [119, 120]. For instance, voxels can be selected if they exhibit a consistent variation in activity across all stimuli [73], by the correlation between real and predicted voxel responses [121], or based on explained variance [122]. Beinborn et al. [74] performed a systematic comparison of voxel selection methods and conclude that explained variance yields the most stable results. Tarhan and Konkle [123] go beyond individual voxels and focus on maximizing the reliability of multi-voxel patterns across voxels with consistent activation across runs.

While most voxel selection methods average over patterns and across participants, Fedorenko et al. [124] proposed an approach to functionally identify language-responsive areas in individual brains without the constraint that these areas need to fall precisely into the same anatomical locations across participants. The regions are selected based on their key signatures of language processing, for example, higher response to sentences than non-word strings. These localized regions can then be probed for their responses to specific experimental manipulations (let's say, the neural activation for abstract and concrete nouns). Participant-specific functional localizers have been shown to increase the sensitivity and functional resolution of fMRI analyses, both in terms of participant aggregation and in the location of language-relevant activations [125]. If the same localizer paradigm is used across individuals, populations, and datasets, the findings can be directly compared to each other [126].

3.3   Available Datasets

In this section, we discuss a small selection of available datasets of cognitive language processing signals to highlight experimental choices that are relevant for analyzing the cognitive plausibility of NLP models. The chosen datasets are recent illustrative cases that all use continuous natural language stimuli (in contrast to controlled, manually constructed phrases or sentences). This is beneficial when using these datasets for NLP since we can investigate language comprehension in the context of naturally occurring texts.

This shift towards naturalistic experiments is becoming more popular in neuroscience and psychology [127]. The largest challenge that comes with using longer continuous inputs is the drastic increase in the duration of the experiment which usually leads to the inclusion of fewer participants. While small samples can be a substantial drawback for neuroscientific analyses, the larger number of words and sentences to train and evaluate our models usually outweighs the smaller number of participants in NLP experiments as long as the data is of high quality.

3.3.1   Annotation Rationales Benchmark

DeYoung et al. [20] combine seven datasets for which human rationales have been collected as supporting evidence for the corresponding labels into the ERASER benchmark. They propose several metrics to capture the alignment between human rationales and computational explanations models. They also measure the faithfulness of computational rationales, i.e., the degree to which the annotated rationales influenced the corresponding label. Such rationales provide valuable information on human decision-making strategies. The individual differences can provide an estimate of the acceptable level of variation for decision patterns in cognitively plausible models.

3.3.2   Self-Paced Reading of Short Stories

The Natural Stories Corpus by Futrell et al. [128] is targeted towards exposing responses to uncommon constructions that are not frequent enough in fully naturalistic corpora, while still maintaining instinctive fluency in the stimuli. It combines naturalistic stimuli from authentic text with manually selected syntactic constructions from controlled psycholinguistic experiments. This is achieved by editing ten English stories from freely available texts to contain an unusual amount of complex low-frequency syntactic constructions while still sounding fluent to native speakers. The texts were presented to 181 native English speakers in a moving window display using a crowd-sourcing platform. The Natural Stories Corpus makes it possible to study how humans rapidly adapt their lexical and syntactic expectations to match the statistics of the current linguistic context [129, 130].

3.3.3   A Multilingual Eye-Tracking Corpus

The majority of cognitively motivated corpora only contain data for a single language (mostly English). The Multilingual Eye-Tracking Corpus (MECO) is a notable exception because it is the first eye-tracking dataset that maintains a stable experimental setup across a broad range of languages. It contains data from 580 participants reading sentences in their first language (L1) and in English (their L2) [131, 132]. The reading material consists of 12 texts describing general-domain topics in approximately 10 sentences. For each text, parallel versions have been created for 13 languages by experts.[1] The complete texts were shown on a single screen spanning multiple lines and the participants read naturally without any time limit.

All participants shared their demographic details and performed cognitive and lexical assessment tests, which allow us to study the impact of user-specific variables such as vocabulary knowledge or age [102]. Siegelman et al. [131] observe that readers of different languages vary considerably in their skipping rate and that this variability is explained by cross-linguistic differences in word length distributions. Such cross-lingual differences are important to consider when evaluating the cognitive plausibility of procedural patterns in Chap. 6.

3.3.4   EEG Datasets of Reading and Listening

We want to highlight two EEG datasets that use continuous stimuli to allow the analysis of authentic language containing words in context rather than focusing on isolated words and manually constructed examples.

The Zurich Cognitive Language Processing Corpus (ZuCo; Hollenstein et al. [91]) is a dataset combining electroencephalography (EEG) and eye-tracking recordings from participants reading natural sentences. ZuCo includes high-density EEG and eye-tracking data of 12 healthy adult native English speakers, each reading natural English text for four to six hours. The recordings span three reading tasks (two related to general understanding and one related to task-specific reading). The reading material was extracted from the Stanford Sentiment Treebank by Socher et al. [133] and the relation extraction corpus by Culotta et al. [134] to reuse the existing annotations for the integration of the cognitive signals with task-specific models.

Ren and Xiong [135] investigate the correlation of ZuCo signals (EEG frequency bands and eye-tracking features) with linguistic information through an attentional mapping procedure. For eye-tracking, they find that the number of re-fixations and the total fixation duration of a word are highly correlated with indicators of lexical difficulty such as polysemy and frequency. For EEG, they observe that power changes in the theta range are more strongly associated with content words than with function words. Differences in syntactic complexity

---

[1] Dutch, English, Estonian, Finnish, German, Greek, Hebrew, Italian, Korean, Norwegian, Russian, Spanish and Turkish.

are reflected in the beta frequency range and the gamma frequency band is sensitive to word order.

Brennan and Hale [136] collected EEG recordings from 52 participants listening to the first chapter of the book *Alice in Wonderland* which consists of 84 sentences [137]. The dataset was recorded to analyze how sequential and hierarchical information guides the expectations that a human listener forms about a word's part of speech when simply listening to everyday language. It has been used to compare the mechanisms of neural grammar models to human syntactic processing [64]. For the same stimuli, fMRI data has also been recorded which enables cross-signal analyses [87, 138].

3.3.5   A Multilingual FMRI Dataset

Li et al. [139] present a very recent example for a multilingual resource which contains multi-echo fMRI data from 49 English, 35 Chinese, and 28 French speakers listening to the audiobook of *Le Petit Prince* in their native language. The corpus has already been preprocessed and contains automated annotations for part-of-speech tags and constituency and dependency parses, as well as manual annotations for co-reference resolution. According to Stehwien et al. [140], it is planned to expand the corpus to a total of 26 languages and to additionally record EEG signals which would yield a highly valuable resource for cross-lingual analyses.

3.4   Ethical Aspects

When collecting a dataset of human responses, it is essential to take on the responsibility of protecting the individual [141]. All collected data comes from humans willing to share their behavioral or physiological data for research expecting to be treated respectfully with high moral standards. It is important to be aware of rules for collecting and storing confidential data, anonymization requirements, and the importance of ethics approval for experimental studies [142]. When planning the experimental setup, we should make sure to follow up-to-date ethical guidelines and avoid reproducing poorly designed experimental choices even if this comes with the cost of impeding comparison with earlier work. For example, it has been shown that text corpora contain recoverable and accurate imprints of our historical biases towards race and gender [143]. With extensive reuse of the same corpora, these biases are propagated to many experiments and affect the conclusions we draw.

As natural language processing researchers, we often re-use existing datasets that have been collected by psychology or neuroscience researchers. This does not free us from considering ethical aspects. Even if we use already anonymized data, we need to make sure that participants cannot be identified based on metadata or idiosyncratic processing patterns that we reveal.

When drawing conclusions based on a dataset, we need to carefully assess the representativeness of the sample. Participant populations in traditional behavioral experiments are usually characterized to be Western, educated, industrialized, rich, and democratic (so-called *WEIRD*, Henrich et al. [144]). This means that the findings about cognitive patterns are extremely biased towards Western culture and it would be presumptuous to expect that findings observed for this group would universally translate to all other demographics. Kidd et al. [145] point out that the extent to which human sentence processing is affected by individual differences is most likely underestimated due to overly homogeneous data samples.

Cognitive signals contain considerable portions of noise and we often oversimplify the signal by aggregating over participants [45, 146]. While aggregation often leads to more robust models, it reduces the signal to a cognitively implausible average that does not correspond anymore to the patterns of any individual. Even if we could derive a prototypical signal, the predictions of such a prototype model would not be sensitive to differences between subgroups. If we abstract from this simplification, we purport a misleading picture of universal cognitive patterns in an idealized human that ignores variation and diversity in human cognition [147]. This leads to overgeneralized conclusions that have serious consequences of demographic exclusion or misrepresentation in applied settings [148]. Recent studies aim at quantifying cross-cultural variation [149, 150], and new culture-sensitive datasets are developed by collecting annotations from fluent speakers instead of propagating labels automatically across languages [151].

A note on language diversity in neurolinguistics

Approximately 7,000 languages from more than 100 language families are spoken and signed across the globe. Language is the only animal communication system that manifests in so many different forms. The cross-linguistic similarities of the language network in the human brain have often been assumed in research. In a study across 45 languages, Malik-Moraleda et al. [152] establish the robustness of the fronto-temporo-parietal language network to cross-linguistic variation as well as the universality of key functional properties, including left-lateralization, strong functional integration among its brain regions and functional selectivity for language processing. Previous cross-lingual studies were limited to two to four languages. For instance, Rueckl et al. [153] examined reading and speech perception using functional MRI in four highly contrasting languages: Spanish, English, Hebrew, and Chinese. They use a range of complementary analyses to show the emergence of common brain signatures across languages independent of the type and depth of their orthographic system.

Eye-tracking and fMRI studies on bilingualism suggest that, although the same general structures are active for both languages, differences within these general structures are present across languages and across levels of processing [120, 154]. For example, the active brain regions for language processing seem to be different for native and foreign language speakers [155] and second language learners exhibit different reading patterns than native speakers [131, 156]. The representation of different languages in bilinguals is influenced by the age of acquisition of the second language (L2), proficiency level of the first language (L1), and the amount of language exposure [157]. In an effort to promote eye-tracking research of bilingual reading, Cop et al. [158] provide an English-Dutch eye-tracking corpus tailored to analyze the bilingual reading process.

While many of the brain activation patterns of the language network in the brain seem to be shared across languages [152], it is important to further explore the differences between language varieties, scripts, and cultures. The limited availability of non-English cognitive data sources restricted research on cognitive plausibility largely to English but new cross-lingual datasets such as the fMRI dataset described in Sect. 3.3.5 or the eye-tracking datasets including L1 and L2 reading from 13 languages in Sect. 3.3 are paving the way for a multilingual perspective.

References

1. Gosse Minnema and Aurélie Herbelot. From brain space to distributional space: The perilous journeys of fMRI decoding. In *Proceedings of the 57th Annual Meeting of the Association for Computational Linguistics: Student Research Workshop*, pages 155–161, Florence, Italy, July 2019. Association for Computational Linguistics. https://doi.org/10.18653/v1/P19-2021. https://aclanthology.org/P19-2021.

2. Nora Hollenstein. *Leveraging Cognitive Processing Signals for Natural Language Understanding*. PhD thesis, ETH Zurich, 2021.

3. James Pustejovsky and Amber Stubbs. *Natural Language Annotation for Machine Learning: A guide to corpus-building for applications*. "O'Reilly Media, Inc.", 2012.

4. Jill Jegerski. Self-paced reading. In *Research methods in second language psycholinguistics*, pages 36–65. Routledge, 2013.

5. Kenneth Holmqvist, Marcus Nyström, Richard Andersson, Richard Dewhurst, Halszka Jarodzka, and Joost Van de Weijer. *Eye tracking: A comprehensive guide to methods and measures*. OUP Oxford, 2011.

6. Mike X Cohen. *Analyzing neural time series data: theory and practice*. MIT press, 2014.

7. Peter Hansen, Morten Kringelbach, and Riitta Salmelin. *MEG: an introduction to methods*. Oxford university press, 2010.

8. José M Soares, Ricardo Magalhães, Pedro S Moreira, Alexandre Sousa, Edward Ganz, Adriana Sampaio, Victor Alves, Paulo Marques, and Nuno Sousa. A hitchhiker's guide to functional magnetic resonance imaging. *Frontiers in neuroscience*, 10:515, 2016.

9. Valentina Quaresima, Silvia Bisconti, and Marco Ferrari. A brief review on the use of functional near-infrared spectroscopy (fnirs) for language imaging studies in human newborns and adults. *Brain and language*, 121(2):79–89, 2012.

10. Graham Wilcock. Introduction to linguistic annotation and text analytics. *Synthesis Lectures on Human Language Technologies*, 2(1):1–159, 2009.

11. Marcel A Just and Patricia A Carpenter. A theory of reading: from eye fixations to comprehension. *Psychological review*, 87(4):329, 1980.

12. Curtis G Northcutt, Anish Athalye, and Jonas Mueller. Pervasive label errors in test sets destabilize machine learning benchmarks. In *Thirty-fifth Conference on Neural Information Processing Systems Datasets and Benchmarks Track (Round 1)*, 2021.

13. Nora Hollenstein, Nathan Schneider, and Bonnie Webber. Inconsistency detection in semantic annotation. In *Proceedings of the Tenth International Conference on Language Resources and Evaluation (LREC'16)*, pages 3986–3990, Portorož, Slovenia, May 2016. European Language Resources Association (ELRA). https://aclanthology.org/L16-1629.

14. Ron Artstein and Massimo Poesio. Inter-coder agreement for computational linguistics. *Computational linguistics*, 34(4):555–596, 2008.

15. David Q. Sun, Hadas Kotek, Christopher Klein, Mayank Gupta, William Li, and Jason D. Williams. Improving human-labeled data through dynamic automatic conflict resolution. In *Proceedings of the 28th International Conference on Computational Linguistics*, pages 3547–3557, Barcelona, Spain (Online), December 2020. International Committee on Computational Linguistics. https://doi.org/10.18653/v1/2020.coling-main.316. https://aclanthology.org/2020.coling-main.316.

16. Jan-Christoph Klie, Bonnie Webber, and Iryna Gurevych. Annotation error detection: Analyzing the past and present for a more coherent future. *arXiv preprint* arXiv:2206.02280, 2022.

17. Jacopo Amidei, Paul Piwek, and Alistair Willis. Rethinking the agreement in human evaluation tasks. In *Proceedings of the 27th International Conference on Computational Linguistics*, pages 3318–3329, Santa Fe, New Mexico, USA, August 2018. Association for Computational Linguistics. https://aclanthology.org/C18-1281.

18. Georg Rasch. Studies in mathematical psychology: I. probabilistic models for some intelligence and attainment tests. 1960.

19. Silviu Paun, Bob Carpenter, Jon Chamberlain, Dirk Hovy, Udo Kruschwitz, and Massimo Poesio. Comparing Bayesian models of annotation. *Transactions of the Association for Computational Linguistics*, 6:571–585, 2018. https://doi.org/10.1162/tacl_a_00040. https://aclanthology.org/Q18-1040.

20. Jay DeYoung, Sarthak Jain, Nazneen Fatema Rajani, Eric Lehman, Caiming Xiong, Richard Socher, and Byron C. Wallace. ERASER: A benchmark to evaluate rationalized NLP models. In *Proceedings of the 58th Annual Meeting of the Association for Computational Linguistics*, pages 4443–4458, Online, July 2020. Association for Computational Linguistics. https://doi.org/10.18653/v1/2020.acl-main.408. https://aclanthology.org/2020.acl-main.408.

21. Pepa Atanasova, Jakob Grue Simonsen, Christina Lioma, and Isabelle Augenstein. A diagnostic study of explainability techniques for text classification. In *Proceedings of the 2020 Conference on Empirical Methods in Natural Language Processing (EMNLP)*, pages 3256–3274, Online, November 2020. Association for Computational Linguistics. https://doi.org/10.18653/v1/2020.emnlp-main.263. https://aclanthology.org/2020.emnlp-main.263.

22. Anders Søgaard. Explainable natural language processing. *Synthesis Lectures on Human Language Technologies*, 14(3):1–123, 2021.

23. Maarten W Van Someren, Yvonne F Barnard, and Jacobijn AC Sandberg. The think aloud method: a practical approach to modelling cognitive. *London: AcademicPress*, 11, 1994.

24. Yiming Zheng, Serena Booth, Julie Shah, and Yilun Zhou. The irrationality of neural rationale models. *arXiv preprint* arXiv:2110.07550, 2021.

25. Amir Bakarov. A survey of word embeddings evaluation methods. *arXiv preprint* arXiv:1801.09536, 2018.

26. Jeremy Auguste, Arnaud Rey, and Benoit Favre. Evaluation of word embeddings against cognitive processes: primed reaction times in lexical decision and naming tasks. In *Proceedings of the 2nd Workshop on Evaluating Vector Space Representations for NLP*, pages 21–26, Copenhagen, Denmark, September 2017. Association for Computational Linguistics. https://doi.org/10.18653/v1/W17-5304. https://aclanthology.org/W17-5304.

27. Wilson L Taylor. "cloze procedure": A new tool for measuring readability. *Journalism quarterly*, 30(4):415–433, 1953.

28. Steven G Luke and Kiel Christianson. The provo corpus: A large eye-tracking corpus with predictability norms. *Behavior research methods*, 50:826–833, 2018.

29. Matthew W Lowder, Wonil Choi, Fernanda Ferreira, and John M Henderson. Lexical predictability during natural reading: Effects of surprisal and entropy reduction. *Cognitive science*, 42:1166–1183, 2018.

30. Kenneth I Forster, Christine Guerrera, and Lisa Elliot. The maze task: Measuring forced incremental sentence processing time. *Behavior research methods*, 41:163–171, 2009.

31. Lisa Beinborn, Torsten Zesch, and Iryna Gurevych. Candidate evaluation strategies for improved difficulty prediction of language tests. In *Proceedings of the Tenth Workshop on Innovative Use of NLP for Building Educational Applications*, pages 1–11, Denver, Colorado, June 2015. Association for Computational Linguistics. https://doi.org/10.3115/v1/W15-0601. https://aclanthology.org/W15-0601.

32. Veronica Boyce, Richard Futrell, and Roger P Levy. Maze made easy: Better and easier measurement of incremental processing difficulty. *Journal of Memory and Language*, 111:104082, 2020.

33. Sian Gooding, Yevgeni Berzak, Tony Mak, and Matt Sharifi. Predicting text readability from scrolling interactions. In *Proceedings of the 25th Conference on Computational Natural Language Learning*, pages 380–390, Online, November 2021. Association for Computational Linguistics. https://doi.org/10.18653/v1/2021.conll-1.30. https://aclanthology.org/2021.conll-1.30.

34. Barbara Plank. Keystroke dynamics as signal for shallow syntactic parsing. In *Proceedings of COLING 2016, the 26th International Conference on Computational Linguistics: Technical Papers*, pages 609–619, Osaka, Japan, December 2016. The COLING 2016 Organizing Committee. https://aclanthology.org/C16-1059.

35. Elma Kerz, Fabio Pruneri, Daniel Wiechmann, Yu Qiao, and Marcus Ströbel. Understanding the dynamics of second language writing through keystroke logging and complexity contours. In *Proceedings of the Twelfth Language Resources and Evaluation Conference*, pages 182–188, Marseille, France, May 2020. European Language Resources Association. ISBN 979-10-95546-34-4. https://aclanthology.org/2020.lrec-1.23.

36. Naoko Witzel, Jeffrey Witzel, and Kenneth Forster. Comparisons of online reading paradigms: Eye tracking, moving-window, and maze. *Journal of psycholinguistic research*, 41(2):105–128, 2012.

37. Stefan L Frank, Irene Fernandez Monsalve, Robin L Thompson, and Gabriella Vigliocco. Reading time data for evaluating broad-coverage models of English sentence processing. *Behavior research methods*, 45(4):1182–1190, 2013.

38. Keith Rayner, Sara C Sereno, Robin K Morris, A Rene Schmauder, and Charles Clifton Jr. Eye movements and on-line language comprehension processes. *Language and Cognitive Processes*, 4(3-4):SI21–SI49, 1989.

39. Benedikt V Ehinger, Katharina Groß, Inga Ibs, and Peter König. A new comprehensive eye-tracking test battery concurrently evaluating the pupil labs glasses and the eyelink 1000. *PeerJ*, 7:e7086, 2019.

40. Javier San Agustin, Henrik Skovsgaard, John Paulin Hansen, and Dan Witzner Hansen. Low-cost gaze interaction: ready to deliver the promises. In *CHI'09 Extended Abstracts on Human Factors in Computing Systems*, pages 4453–4458, 2009.

41. Jose Gómez-Poveda and Elena Gaudioso. Evaluation of temporal stability of eye tracking algorithms using webcams. *Expert Systems with Applications*, 64:69–83, 2016.

42. Alexandra Papoutsaki, Patsorn Sangkloy, James Laskey, Nediyana Daskalova, Jeff Huang, and James Hays. Webgazer: Scalable webcam eye tracking using user interactions. In *Proceedings of the 25th International Joint Conference on Artificial Intelligence (IJCAI)*, pages 3839–3845. AAAI, 2016.

43. Xiu Guan, Chaojing Lei, Yingfen Huang, Yu Chen, Hanyue Du, Shuowen Zhang, and Xiang Feng. An analysis of reading process based on real-time eye-tracking data with web-camera–focus on english reading at higher education level. In *Proceedings of the 4th Workshop on Predicting Performance Based on the Analysis of Reading Behavior*, 2022.

44. Weston Sewell and Oleg Komogortsev. Real-time eye gaze tracking with an unmodified commodity webcam employing a neural network. In *CHI' 10 Extended Abstracts on Human Factors in Computing Systems*, pages 3739–3744, 2010.

45. Nora Hollenstein, Maria Barrett, and Lisa Beinborn. Towards best practices for leveraging human language processing signals for natural language processing. In *Proceedings of the Second Workshop on Linguistic and Neurocognitive Resources*, pages 15–27, Marseille, France, May 2020. European Language Resources Association. ISBN 979-10-95546-52-8. https://aclanthology.org/2020.lincr-1.3.

46. Vera Demberg and Frank Keller. Data from eye-tracking corpora as evidence for theories of syntactic processing complexity. *Cognition*, 109(2):193–210, 2008.

47. Leila Wehbe, Idan Asher Blank, Cory Shain, Richard Futrell, Roger Levy, Titus von der Malsburg, Nathaniel Smith, Edward Gibson, and Evelina Fedorenko. Incremental language comprehension difficulty predicts activity in the language network but not the multiple demand network. *Cerebral Cortex*, 31(9):4006–4023, 2021.

48. Maria Barrett, Joachim Bingel, Frank Keller, and Anders Søgaard. Weakly supervised part-of-speech tagging using eye-tracking data. In *Proceedings of the 54th Annual Meeting of the Association for Computational Linguistics (Volume 2: Short Papers)*, pages 579–584, Berlin, Germany, August 2016. Association for Computational Linguistics. https://doi.org/10.18653/v1/P16-2094. https://aclanthology.org/P16-2094.

49. Michalina Strzyz, David Vilares, and Carlos Gómez-Rodríguez. Towards making a dependency parser see. In *Proceedings of the 2019 Conference on Empirical Methods in Natural Language Processing and the 9th International Joint Conference on Natural Language Processing (EMNLP-IJCNLP)*, pages 1500–1506, Hong Kong, China, November 2019. Association for Computational Linguistics. https://doi.org/10.18653/v1/D19-1160. https://aclanthology.org/D19-1160.

50. Abhijit Mishra, Diptesh Kanojia, Seema Nagar, Kuntal Dey, and Pushpak Bhattacharyya. Leveraging cognitive features for sentiment analysis. In *Proceedings of the 20th SIGNLL Conference on Computational Natural Language Learning*, pages 156–166, Berlin, Germany, August 2016. Association for Computational Linguistics. https://doi.org/10.18653/v1/K16-1016. https://aclanthology.org/K16-1016.

51. Duo Yang and Nora Hollenstein. Plm-as: Pre-trained language models augmented with scanpaths for sentiment classification. In *Proceedings of the Northern Lights Deep Learning Workshop*, volume 4, 2023.

52. Varun Khurana, Yaman Kumar, Nora Hollenstein, Rajesh Kumar, and Balaji Krishnamurthy. Synthesizing human gaze feedback for improved NLP performance. In *Proceedings of the 17th Conference of the European Chapter of the Association for Computational Linguistics*, pages 1887–1900, Dubrovnik, Croatia, May 2023. Association for Computational Linguistics. https://aclanthology.org/2023.eacl-main.139.

53. Neeru Dubey, Simran Setia, Amit Arjun Verma, and SRS Iyengar. Wikigaze: Gaze-based personalized summarization of wikipedia reading session. In *Proceedings of the 3rd Workshop on Human Factors in Hypertext*, pages 1–9, 2020.

54. Anna M Beres. Time is of the essence: A review of electroencephalography (eeg) and event-related brain potentials (erps) in language research. *Applied psychophysiology and biofeedback*, 42(4):247–255, 2017.

55. Barbara Penolazzi, Alessandro Angrilli, and Remo Job. Gamma EEG activity induced by semantic violation during sentence reading. *Neuroscience Letters*, 465(1):74–78, 2009.
56. Arnaud Delorme and Scott Makeig. Eeglab: an open source toolbox for analysis of single-trial eeg dynamics including independent component analysis. *Journal of neuroscience methods*, 134(1):9–21, 2004.
57. Andreas Pedroni, Amirreza Bahreini, and Nicolas Langer. Automagic: Standardized preprocessing of big eeg data. *NeuroImage*, 200:460–473, 2019.
58. Tamara Y Swaab, Kerry Ledoux, C Christine Camblin, and Megan A Boudewyn. Language-related erp components. *Oxford handbook of event-related potential components*, pages 397–440, 2012.
59. Marta Kutas and Steven A Hillyard. Reading between the lines: Event-related brain potentials during natural sentence processing. *Brain and language*, 11(2):354–373, 1980.
60. Steven G Luke and Kiel Christianson. Limits on lexical prediction during reading. *Cognitive Psychology*, 88:22–60, 2016.
61. Katherine A DeLong and Marta Kutas. Comprehending surprising sentences: sensitivity of post-n400 positivities to contextual congruity and semantic relatedness. *Language, Cognition and Neuroscience*, 35(8):1044–1063, 2020.
62. Gina R Kuperberg, Trevor Brothers, and Edward W Wlotko. A tale of two positivities and the n400: Distinct neural signatures are evoked by confirmed and violated predictions at different levels of representation. *Journal of Cognitive Neuroscience*, 32(1):12–35, 2020.
63. James A Michaelov, Seana Coulson, and Benjamin K Bergen. So cloze yet so far: N400 amplitude is better predicted by distributional information than human predictability judgements. *IEEE Transactions on Cognitive and Developmental Systems*, 2022.
64. John Hale, Chris Dyer, Adhiguna Kuncoro, and Jonathan Brennan. Finding syntax in human encephalography with beam search. In *Proceedings of the 56th Annual Meeting of the Association for Computational Linguistics (Volume 1: Long Papers)*, pages 2727–2736, Melbourne, Australia, July 2018. Association for Computational Linguistics. https://doi.org/10.18653/v1/P18-1254. https://aclanthology.org/P18-1254.
65. Danny Merkx and Stefan L. Frank. Human sentence processing: Recurrence or attention? In *Proceedings of the Workshop on Cognitive Modeling and Computational Linguistics*, pages 12–22, Online, June 2021. Association for Computational Linguistics. https://doi.org/10.18653/v1/2021.cmcl-1.2. https://aclanthology.org/2021.cmcl-1.2.
66. Allyson Ettinger. What BERT is not: Lessons from a new suite of psycholinguistic diagnostics for language models. *Transactions of the Association for Computational Linguistics*, 8:34–48, 2020. https://doi.org/10.1162/tacl_a_00298. https://aclanthology.org/2020.tacl-1.3.
67. Pengfei Sun, Gopala K Anumanchipalli, and Edward F Chang. Brain2char: A deep architecture for decoding text from brain recordings. *Journal of Neural Engineering*, 2020.
68. Nicolas Affolter, Beni Egressy, Damian Pascual, and Roger Wattenhofer. Brain2word: Decoding brain activity for language generation. *arXiv preprint* arXiv:2009.04765, 2020.
69. Christian Pfeiffer, Nora Hollenstein, Ce Zhang, and Nicolas Langer. Neural dynamics of sentiment processing during naturalistic sentence reading. *NeuroImage*, page 116934, 2020.
70. Oscar Esteban, Christopher J Markiewicz, Ross W Blair, Craig A Moodie, A Ilkay Isik, Asier Erramuzpe, James D Kent, Mathias Goncalves, Elizabeth DuPre, Madeleine Snyder, et al. fmriprep: a robust preprocessing pipeline for functional mri. *Nature methods*, 16(1):111–116, 2019.
71. Francis M Miezin, L Maccotta, JM Ollinger, SE Petersen, and RL Buckner. Characterizing the hemodynamic response: effects of presentation rate, sampling procedure, and the possibility of ordering brain activity based on relative timing. *Neuroimage*, 11(6):735–759, 2000.

72. Alexander G Huth, Wendy A De Heer, Thomas L Griffiths, Frédéric E Theunissen, and Jack L Gallant. Natural speech reveals the semantic maps that tile human cerebral cortex. *Nature*, 532(7600):453–458, 2016.

73. Tom M Mitchell, Svetlana V Shinkareva, Andrew Carlson, Kai-Min Chang, Vicente L Malave, Robert A Mason, and Marcel Adam Just. Predicting human brain activity associated with the meanings of nouns. *science*, 320(5880):1191–1195, 2008.

74. Lisa Beinborn, Samira Abnar, and Rochelle Choenni. Robust evaluation of language-brain encoding experiments. *International Journal of Computational Linguistics and Applications*, 2019.

75. Colette M McKay, Adnan Shah, Abd-Krim Seghouane, Xin Zhou, William Cross, and Ruth Litovsky. Connectivity in language areas of the brain in cochlear implant users as revealed by fnirs. In *Physiology, psychoacoustics and cognition in normal and impaired hearing*, pages 327–335. Springer, Cham, 2016.

76. Lilian Cristine Scherer, Rochele Paz Fonseca, Mahnoush Amiri, Daniel Adrover-Roig, Karine Marcotte, Francine Giroux, Noureddine Senhadji, Habib Benali, Frédéric Lesage, and Ana Inés Ansaldo. Syntactic processing in bilinguals: An fnirs study. *Brain and language*, 121(2):144–151, 2012.

77. Lu Cao, Dandan Huang, Yue Zhang, Xiaowei Jiang, and Yanan Chen. Brain decoding using fNIRS. In *Proceedings of the AAAI Conference on Artificial Intelligence*, volume 35, pages 12602–12611, 2021.

78. Licui Zhao, Haruyuki Kojima, Daichi Yasunaga, and Koji Irie. Syntactic and semantic processing in japanese sentence reading: A research using functional near-infrared spectroscopy (fnirs). *Journal of Psycholinguistic Research*, pages 1–17, 2021.

79. Nora Hollenstein and Ce Zhang. Entity recognition at first sight: Improving NER with eye movement information. In *Proceedings of the 2019 Conference of the North American Chapter of the Association for Computational Linguistics: Human Language Technologies, Volume 1 (Long and Short Papers)*, pages 1–10, Minneapolis, Minnesota, June 2019. Association for Computational Linguistics. https://doi.org/10.18653/v1/N19-1001. https://aclanthology.org/N19-1001.

80. Arthur Mensch, Julien Mairal, Danilo Bzdok, Bertrand Thirion, and Gaël Varoquaux. Learning neural representations of human cognition across many fmri studies. *Advances in neural information processing systems*, 30, 2017.

81. Dan Schwartz, Mariya Toneva, and Leila Wehbe. Inducing brain-relevant bias in natural language processing models. *Advances in neural information processing systems*, 32, 2019.

82. Nora Hollenstein, Antonio de la Torre, Nicolas Langer, and Ce Zhang. CogniVal: A framework for cognitive word embedding evaluation. In *Proceedings of the 23rd Conference on Computational Natural Language Learning (CoNLL)*, pages 538–549, Hong Kong, China, November 2019. Association for Computational Linguistics. https://doi.org/10.18653/v1/K19-1050. https://aclanthology.org/K19-1050.

83. Maria Barrett, Ana Valeria González-Garduño, Lea Frermann, and Anders Søgaard. Unsupervised induction of linguistic categories with records of reading, speaking, and writing. In *Proceedings of the 2018 Conference of the North American Chapter of the Association for Computational Linguistics: Human Language Technologies, Volume 1 (Long Papers)*, pages 2028–2038, New Orleans, Louisiana, June 2018. Association for Computational Linguistics. https://doi.org/10.18653/v1/N18-1184. https://aclanthology.org/N18-1184.

84. Stefan L Frank, Leun J Otten, Giulia Galli, and Gabriella Vigliocco. The ERP response to the amount of information conveyed by words in sentences. *Brain and language*, 140:1–11, 2015.

85. Richard Futrell, Edward Gibson, Harry J Tily, Idan Blank, Anastasia Vishnevetsky, Steven T Piantadosi, and Evelina Fedorenko. The natural stories corpus: a reading-time corpus of english

texts containing rare syntactic constructions. *Language Resources and Evaluation*, 55(1):63–77, 2021.

86. Cory Shain, Idan Asher Blank, Marten van Schijndel, William Schuler, and Evelina Fedorenko. fmri reveals language-specific predictive coding during naturalistic sentence comprehension. *Neuropsychologia*, 138:107307, 2020.

87. Jonathan R Brennan, Edward P Stabler, Sarah E Van Wagenen, Wen-Ming Luh, and John T Hale. Abstract linguistic structure correlates with temporal activity during naturalistic comprehension. *Brain and language*, 157:81–94, 2016.

88. Christoph Mulert. Simultaneous eeg and fmri: towards the characterization of structure and dynamics of brain networks. *Dialogues in clinical neuroscience*, 2022.

89. Olaf Dimigen, Werner Sommer, Annette Hohlfeld, Arthur M Jacobs, and Reinhold Kliegl. Coregistration of eye movements and eeg in natural reading: analyses and review. *Journal of experimental psychology: General*, 140(4):552, 2011.

90. John M Henderson, Steven G Luke, Joseph Schmidt, and John E Richards. Co-registration of eye movements and event-related potentials in connected-text paragraph reading. *Frontiers in systems neuroscience*, 7:28, 2013.

91. Nora Hollenstein, Jonathan Rotsztejn, Marius Troendle, Andreas Pedroni, Ce Zhang, and Nicolas Langer. Zuco, a simultaneous eeg and eye-tracking resource for natural sentence reading. *Scientific data*, 5(1):1–13, 2018.

92. John M Henderson, Wonil Choi, Matthew W Lowder, and Fernanda Ferreira. Language structure in the brain: A fixation-related fmri study of syntactic surprisal in reading. *Neuroimage*, 132:293–300, 2016.

93. Yun-Nung Chen, Kai-min Kevin Chang, and Jack Mostow. Towards using eeg to improve asr accuracy. In *Proceedings of the 2012 Conference of the North American Chapter of the Association for Computational Linguistics: Human Language Technologies*, pages 382–385, 2012.

94. Zhenhailong Wang and Heng Ji. Open vocabulary electroencephalography-to-text decoding and zero-shot sentiment classification. In *Proceedings of the AAAI Conference on Artificial Intelligence*, volume 36, pages 5350–5358, 2022.

95. Nora Hollenstein, Cedric Renggli, Benjamin Glaus, Maria Barrett, Marius Troendle, Nicolas Langer, and Ce Zhang. Decoding eeg brain activity for multi-modal natural language processing. *Frontiers in Human Neuroscience*, page 378, 2021.

96. Mariya Toneva and Leila Wehbe. Interpreting and improving natural-language processing (in machines) with natural language-processing (in the brain). *Advances in Neural Information Processing Systems*, 32, 2019.

97. Clark Glymour and Catherine Hanson. Reverse inference in neuropsychology. *The British Journal for the Philosophy of Science*, 2016.

98. Nora Hollenstein, Maria Barrett, Marius Troendle, Francesco Bigiolli, Nicolas Langer, and Ce Zhang. Advancing nlp with cognitive language processing signals. *arXiv preprint* arXiv:1904.02682, 2019.

99. Nora Hollenstein, Marius Tröndle, Martyna Plomecka, Samuel Kiegeland, Yilmazcan Özyurt, Lena A Jäger, and Nicolas Langer. The zuco benchmark on cross-subject reading task classification with eeg and eye-tracking data. *bioRxiv*, 2022.

100. Ekaterina Artemova, Amir Bakarov, Aleksey Artemov, Evgeny Burnaev, and Maxim Sharaev. Data-driven models and computational tools for neurolinguistics: a language technology perspective. *Journal of Cognitive Science*, 21(1):15–52, 2020.

101. Holger Schielzeth, Niels J Dingemanse, Shinichi Nakagawa, David F Westneat, Hassen Allegue, Céline Teplitsky, Denis Réale, Ned A Dochtermann, László Zsolt Garamszegi, and Yimen G

Araya-Ajoy. Robustness of linear mixed-effects models to violations of distributional assumptions. *Methods in ecology and evolution*, 11(9):1141–1152, 2020.

102. Stephanie Brandl and Nora Hollenstein. Every word counts: A multilingual analysis of individual human alignment with model attention. In *Proceedings of the 2nd Conference of the Asia-Pacific Chapter of the Association for Computational Linguistics and the 12th International Joint Conference on Natural Language Processing (Volume 2: Short Papers)*, pages 72–77, Online only, November 2022. Association for Computational Linguistics. https://aclanthology.org/2022.aacl-short.10.

103. Dustin Scheinost, Stephanie Noble, Corey Horien, Abigail S Greene, Evelyn MR Lake, Mehraveh Salehi, Siyuan Gao, Xilin Shen, David O'Connor, Daniel S Barron, et al. Ten simple rules for predictive modeling of individual differences in neuroimaging. *NeuroImage*, 193:35–45, 2019.

104. Nora Hollenstein, Federico Pirovano, Ce Zhang, Lena Jäger, and Lisa Beinborn. Multilingual language models predict human reading behavior. In *Proceedings of the 2021 Conference of the North American Chapter of the Association for Computational Linguistics: Human Language Technologies*, pages 106–123, Online, June 2021. Association for Computational Linguistics. https://doi.org/10.18653/v1/2021.naacl-main.10. https://aclanthology.org/2021.naacl-main.10.

105. Nora Hollenstein and Lisa Beinborn. Relative importance in sentence processing. In *Proceedings of the 59th Annual Meeting of the Association for Computational Linguistics and the 11th International Joint Conference on Natural Language Processing (Volume 2: Short Papers)*, pages 141–150, Online, August 2021. Association for Computational Linguistics. https://doi.org/10.18653/v1/2021.acl-short.19. https://aclanthology.org/2021.acl-short.19.

106. Ana Gonzalez-Garduno and Anders Søgaard. Learning to predict readability using eye-movement data from natives and learners. In *Proceedings of the AAAI Conference on Artificial Intelligence*, volume 32, 2018.

107. Marcel A Just, Patricia A Carpenter, and Jacqueline D Woolley. Paradigms and processes in reading comprehension. *Journal of experimental psychology: General*, 111(2):228, 1982.

108. Shravan Vasishth and Richard L Lewis. Argument-head distance and processing complexity: Explaining both locality and antilocality effects. *Language*, pages 767–794, 2006.

109. Nathaniel J Smith and Roger Levy. The effect of word predictability on reading time is logarithmic. *Cognition*, 128(3):302–319, 2013.

110. Shravan Vasishth. On the proper treatment of spillover in real-time reading studies: Consequences for psycholinguistic theories. In *Proceedings of the international conference on linguistic evidence*, pages 96–100, 2006.

111. Cory Shain and William Schuler. Continuous-time deconvolutional regression for psycholinguistic modeling. *Cognition*, 215:104735, 2021.

112. Leila Wehbe, Brian Murphy, Partha Talukdar, Alona Fyshe, Aaditya Ramdas, and Tom Mitchell. Simultaneously uncovering the patterns of brain regions involved in different story reading subprocesses. *PloS one*, 9(11):e112575, 2014.

113. Shailee Jain, Vy Vo, Shivangi Mahto, Amanda LeBel, Javier S Turek, and Alexander Huth. Interpretable multi-timescale models for predicting fmri responses to continuous natural speech. *Advances in Neural Information Processing Systems*, 33:13738–13749, 2020.

114. Samira Abnar, Lisa Beinborn, Rochelle Choenni, and Willem Zuidema. Blackbox meets blackbox: Representational similarity & stability analysis of neural language models and brains. In *Proceedings of the 2019 ACL Workshop BlackboxNLP: Analyzing and Interpreting Neural Networks for NLP*, pages 191–203, Florence, Italy, August 2019. Association for Computational Linguistics. https://doi.org/10.18653/v1/W19-4820. https://aclanthology.org/W19-4820.

115. Xinyang Yu, Pharino Chum, and Kwee-Bo Sim. Analysis the effect of pca for feature reduction in non-stationary eeg based motor imagery of bci system. *Optik*, 125(3):1498–1502, 2014.
116. Abdulhamit Subasi and M Ismail Gursoy. Eeg signal classification using pca, ica, lda and support vector machines. *Expert systems with applications*, 37(12):8659–8666, 2010.
117. Jose Antonio Urigüen and Begoña Garcia-Zapirain. Eeg artifact removal-state-of-the-art and guidelines. *Journal of neural engineering*, 12(3):031001, 2015.
118. Olaf Dimigen. Optimizing the ica-based removal of ocular eeg artifacts from free viewing experiments. *NeuroImage*, 207:116117, 2020.
119. Nikolaus Kriegeskorte, Rainer Goebel, and Peter Bandettini. Information-based functional brain mapping. *Proceedings of the National Academy of Sciences*, 103(10):3863–3868, 2006.
120. Morteza Dehghani, Reihane Boghrati, Kingson Man, Joe Hoover, Sarah I Gimbel, Ashish Vaswani, Jason D Zevin, Mary Helen Immordino-Yang, Andrew S Gordon, Antonio Damasio, et al. Decoding the neural representation of story meanings across languages. *Human brain mapping*, 38(12):6096–6106, 2017.
121. Shailee Jain and Alexander Huth. Incorporating context into language encoding models for fmri. *Advances in neural information processing systems*, 31, 2018.
122. Jon Gauthier and Anna Ivanova. Does the brain represent words? an evaluation of brain decoding studies of language understanding. *arXiv preprint* arXiv:1806.00591, 2018.
123. Leyla Tarhan and Talia Konkle. Reliability-based voxel selection. *NeuroImage*, 207:116350, 2020.
124. Evelina Fedorenko, Po-Jang Hsieh, Alfonso Nieto-Castañón, Susan Whitfield-Gabrieli, and Nancy Kanwisher. New method for fmri investigations of language: defining rois functionally in individual subjects. *Journal of neurophysiology*, 104(2):1177–1194, 2010.
125. Alfonso Nieto-Castañón and Evelina Fedorenko. Subject-specific functional localizers increase sensitivity and functional resolution of multi-subject analyses. *Neuroimage*, 63(3):1646–1669, 2012.
126. Evelina Fedorenko, Idan Asher Blank, Matthew Siegelman, and Zachary Mineroff. Lack of selectivity for syntax relative to word meanings throughout the language network. *Cognition*, 203:104348, 2020.
127. Phillip M Alday. M/eeg analysis of naturalistic stories: a review from speech to language processing. *Language, Cognition and Neuroscience*, 34(4):457–473, 2019.
128. Richard Futrell, Edward Gibson, Harry J. Tily, Idan Blank, Anastasia Vishnevetsky, Steven Piantadosi, and Evelina Fedorenko. The natural stories corpus. In *Proceedings of the Eleventh International Conference on Language Resources and Evaluation (LREC 2018)*, Miyazaki, Japan, May 2018. European Language Resources Association (ELRA). https://aclanthology.org/L18-1012.
129. Marten van Schijndel and Tal Linzen. A neural model of adaptation in reading. In *Proceedings of the 2018 Conference on Empirical Methods in Natural Language Processing*, pages 4704–4710, Brussels, Belgium, October-November 2018. Association for Computational Linguistics. https://doi.org/10.18653/v1/D18-1499. https://aclanthology.org/D18-1499.
130. Cory Shain. CDRNN: Discovering complex dynamics in human language processing. In *Proceedings of the 59th Annual Meeting of the Association for Computational Linguistics and the 11th International Joint Conference on Natural Language Processing (Volume 1: Long Papers)*, pages 3718–3734, Online, August 2021. Association for Computational Linguistics. https://doi.org/10.18653/v1/2021.acl-long.288. https://aclanthology.org/2021.acl-long.288.
131. Noam Siegelman, Sascha Schroeder, Cengiz Acartürk, Hee-Don Ahn, Svetlana Alexeeva, Simona Amenta, Raymond Bertram, Rolando Bonandrini, Marc Brysbaert, Daria Chernova, et al. Expanding horizons of cross-linguistic research on reading: The multilingual eye-movement corpus (meco). *Behavior research methods*, pages 1–21, 2022.

132. Victor Kuperman, Noam Siegelman, Sascha Schroeder, Cengiz Acartürk, Svetlana Alexeeva, Simona Amenta, Raymond Bertram, Rolando Bonandrini, Marc Brysbaert, Daria Chernova, et al. Text reading in english as a second language: Evidence from the multilingual eye-movements corpus. *Studies in Second Language Acquisition*, pages 1–35, 2022.

133. Richard Socher, Alex Perelygin, Jean Wu, Jason Chuang, Christopher D. Manning, Andrew Ng, and Christopher Potts. Recursive deep models for semantic compositionality over a sentiment treebank. In *Proceedings of the 2013 Conference on Empirical Methods in Natural Language Processing*, pages 1631–1642, Seattle, Washington, USA, October 2013. Association for Computational Linguistics. https://aclanthology.org/D13-1170.

134. Aron Culotta, Andrew McCallum, and Jonathan Betz. Integrating probabilistic extraction models and data mining to discover relations and patterns in text. In *Proceedings of the Human Language Technology Conference of the NAACL, Main Conference*, pages 296–303, New York City, USA, June 2006. Association for Computational Linguistics. https://aclanthology.org/N06-1038.

135. Yuqi Ren and Deyi Xiong. Bridging between cognitive processing signals and linguistic features via a unified attentional network. In *Proceedings of the AAAI Conference on Artificial Intelligence*, volume 36, pages 49–58, 2022.

136. Jonathan R Brennan and John T Hale. Hierarchical structure guides rapid linguistic predictions during naturalistic listening. *PloS one*, 14(1):e0207741, 2019.

137. Lewis Carroll. *Alice's adventures in Wonderland*. Macmillan, 1865.

138. Shohini Bhattasali, Jonathan Brennan, Wen-Ming Luh, Berta Franzluebbers, and John Hale. The alice datasets: fMRI & EEG observations of natural language comprehension. In *Proceedings of the Twelfth Language Resources and Evaluation Conference*, pages 120–125, Marseille, France, May 2020. European Language Resources Association. ISBN 979-10-95546-34-4. https://aclanthology.org/2020.lrec-1.15.

139. Jixing Li, Shohini Bhattasali, Shulin Zhang, Berta Franzluebbers, Wen-Ming Luh, R Nathan Spreng, Jonathan R Brennan, Yiming Yang, Christophe Pallier, and John Hale. Le petit prince multilingual naturalistic fmri corpus. *Scientific data*, 9(1):1–15, 2022.

140. Sabrina Stehwien, Lena Henke, John Hale, Jonathan Brennan, and Lars Meyer. The little prince in 26 languages: Towards a multilingual neuro-cognitive corpus. In *Proceedings of the Second Workshop on Linguistic and Neurocognitive Resources*, pages 43–49, Marseille, France, May 2020. European Language Resources Association. ISBN 979-10-95546-52-8. https://aclanthology.org/2020.lincr-1.6.

141. Simon Šuster, Stéphan Tulkens, and Walter Daelemans. A short review of ethical challenges in clinical natural language processing. *arXiv preprint* arXiv:1703.10090, 2017.

142. Emily M. Bender and Batya Friedman. Data statements for natural language processing: Toward mitigating system bias and enabling better science. *Transactions of the Association for Computational Linguistics*, 6:587–604, 2018. https://doi.org/10.1162/tacl_a_00041. https://aclanthology.org/Q18-1041.

143. Aylin Caliskan, Joanna J Bryson, and Arvind Narayanan. Semantics derived automatically from language corpora contain human-like biases. *Science*, 356(6334):183–186, 2017.

144. Joseph Henrich, Steven J Heine, and Ara Norenzayan. The weirdest people in the world? *Behavioral and brain sciences*, 33(2-3):61–83, 2010.

145. Evan Kidd, Seamus Donnelly, and Morten H Christiansen. Individual differences in language acquisition and processing. *Trends in cognitive sciences*, 22(2):154–169, 2018.

146. Sigrid Klerke and Barbara Plank. At a glance: The impact of gaze aggregation views on syntactic tagging. In *Proceedings of the Beyond Vision and LANguage: inTEgrating Real-world kNowledge (LANTERN)*, pages 51–61, Hong Kong, China, November 2019. Association for Computational Linguistics. https://doi.org/10.18653/v1/D19-6408. https://aclanthology.org/D19-6408.

147. Stephen C Levinson. The original sin of cognitive science. *Topics in cognitive science*, 4(3):396–403, 2012.
148. Dirk Hovy and Shannon L. Spruit. The social impact of natural language processing. In *Proceedings of the 54th Annual Meeting of the Association for Computational Linguistics (Volume 2: Short Papers)*, pages 591–598, Berlin, Germany, August 2016. Association for Computational Linguistics. https://doi.org/10.18653/v1/P16-2096. https://aclanthology.org/P16-2096.
149. Michael Muthukrishna, Adrian V Bell, Joseph Henrich, Cameron M Curtin, Alexander Gedranovich, Jason McInerney, and Braden Thue. Beyond western, educated, industrial, rich, and democratic (weird) psychology: Measuring and mapping scales of cultural and psychological distance. *Psychological science*, 31(6):678–701, 2020.
150. Edmond Awad, Sohan Dsouza, Richard Kim, Jonathan Schulz, Joseph Henrich, Azim Shariff, Jean-François Bonnefon, and Iyad Rahwan. The moral machine experiment. *Nature*, 563(7729):59–64, 2018.
151. Fangyu Liu, Emanuele Bugliarello, Edoardo Maria Ponti, Siva Reddy, Nigel Collier, and Desmond Elliott. Visually grounded reasoning across languages and cultures. In *Proceedings of the 2021 Conference on Empirical Methods in Natural Language Processing*, pages 10467–10485, Online and Punta Cana, Dominican Republic, November 2021. Association for Computational Linguistics. https://doi.org/10.18653/v1/2021.emnlp-main.818. https://aclanthology.org/2021.emnlp-main.818.
152. Saima Malik-Moraleda, Dima Ayyash, Jeanne Gallée, Josef Affourtit, Malte Hoffmann, Zachary Mineroff, Olessia Jouravlev, and Evelina Fedorenko. An investigation across 45 languages and 12 language families reveals a universal language network. *Nature Neuroscience*, 25(8):1014–1019, 2022.
153. Jay G Rueckl, Pedro M Paz-Alonso, Peter J Molfese, Wen-Jui Kuo, Atira Bick, Stephen J Frost, Roeland Hancock, Denise H Wu, William Einar Mencl, Jon Andoni Duñabeitia, et al. Universal brain signature of proficient reading: Evidence from four contrasting languages. *Proceedings of the National Academy of Sciences*, 112(50):15510–15515, 2015.
154. Viorica Marian, Michael Spivey, and Joy Hirsch. Shared and separate systems in bilingual language processing: Converging evidence from eyetracking and brain imaging. *Brain and language*, 86(1):70–82, 2003.
155. Daniela Perani, Stanislas Dehaene, Franco Grassi, Laurent Cohen, Stefano F Cappa, Emmanuel Dupoux, Ferruccio Fazio, and Jacques Mehler. Brain processing of native and foreign languages. *NeuroReport-International Journal for Rapid Communications of Research in Neuroscience*, 7(15):2439–2444, 1996.
156. Paola E Dussias. Uses of eye-tracking data in second language sentence processing research. *Annual Review of Applied Linguistics*, 30:149–166, 2010.
157. Monika M Połczyńska and Susan Y Bookheimer. General principles governing the amount of neuroanatomical overlap between languages in bilinguals. *Neuroscience & Biobehavioral Reviews*, 130:1–14, 2021.
158. Uschi Cop, Nicolas Dirix, Denis Drieghe, and Wouter Duyck. Presenting GECO: An eyetracking corpus of monolingual and bilingual sentence reading. *Behavior Research Methods*, 49(2):602–615, 2017.

Behavioral Patterns

4

It is difficult for humans to understand and interpret the learning processes in neural networks because our brain is not well equipped for interpreting abstract matrix transformations in high-dimensional space [1]. Interpretability methods and visualizations can help us simplify the internal processes of natural language processing models and conceptualize them in a more accessible way. But even if we consider neural models as black boxes, we can understand more about their decision processes by analyzing behavioral patterns of input–output pairs.

In psychology, behavioral experiments with humans have been conducted for more than a century [2]. When conducting experiments with computational models, we have the advantage that we have to account for fewer factors of variability: trained models are usually deterministic, i.e., they always return the same result for the same input,[1] and their performance is not influenced by external factors. Human responses, in contrast, can vary depending on the participants' opinions, motivation, and concentration, as well as physical factors such as hunger or fatigue, or even environmental factors such as light or sound conditions in the lab. This advantage comes with the responsibility of methodological rigor to avoid falling into the trap of cherry-picking results from multiple comparisons.

In this chapter, we discuss how behavioral patterns in NLP models can be analyzed. We propose that integrating the concept of instance difficulty is more relevant from a cognitive perspective than the average performance across all instances. In practice, the average performance of current NLP models often outperforms human annotators in specific tasks [3]. Nevertheless, models tend to fail drastically on out-of-domain instances that pose no problem for humans [4]. These findings indicate that certain language processing abilities that are crucial for generalization are not acquired by models [5]. We discuss how models can be systematically tested for linguistic knowledge and introduce methods to diagnose cognitively implausible sensitivity against perturbations. We argue that fine-grained evalua-

---

[1] Unless dropout is applied during testing, see Sect. 4.1.2.

© The Author(s), under exclusive license to Springer Nature Switzerland AG 2024
L. Beinborn and N. Hollenstein, *Cognitive Plausibility in Natural Language Processing*, Synthesis Lectures on Human Language Technologies, https://doi.org/10.1007/978-3-031-43260-6_4

tions on the instance level are required to assess the cognitive plausibility of models and that it is crucial to adopt a multilingual perspective. We are optimistic that cognitively informed curriculum learning can increase the generalizability of models without resorting to ever larger datasets.

4.1   Analyzing Behavioral Patterns

We first revisit established techniques for data and error analysis from a cognitive perspective and then expand on the concept of difficulty.

4.1.1   Data and Error Analysis

The value of the obvious first step of examining neural models is often underestimated: analyzing the data that comes in and the data that comes out. The input data determine the information the model can access during learning, and the model outputs provide insight into the patterns that the model learned.

4.1.1.1 Input Characteristics

In psychological studies of language processing, input stimuli are usually carefully designed to control for any confounding factors. In natural language processing, we generally try to use more naturalistic data to make the models more robust to the noise that can be expected in application scenarios. Large pre-trained language models are intentionally trained on a variety of genres ranging from carefully curated newspaper texts over fictional childrens' books to casual forum posts [6]. The size of the training data is so large that it becomes infeasible to analyze the whole dataset. For the PALM language model, only 1% of the dataset was analyzed using descriptive statistics to detect the existence of person-identifying information and biases with respect to gender, race, and religion [7]. These data-specific biases are propagated to the performance of the model and can lead to exclusion or demographic misrepresentation [8]. Being unaware of the extent of such biases in the training data makes it impossible to develop balancing strategies in applied scenarios. Disappointingly, the analysis of commercially used pre-training data commonly remains infeasible within the academic infrastructure because of a lack of information on the composition and filtering of (potentially proprietary) data and the requirement of excessive computing resources. Only the task-specific training data for the finetuning phase can usually be inspected and analyzed more thoroughly.

When analyzing the input data for supervised scenarios, two aspects play a role: the characteristics of the data itself and the characteristics of the annotations. In order to identify the peculiarities of the data, we explore its content characteristics on multiple linguistic processing levels and its metadata statistics (e.g., sources, number of authors, geographical origin, language) and compare the results to other data samples. The validity of the annotations can be evaluated by analyzing the number of annotators and their backgrounds, the underlying rationale in the annotation guidelines, and the strategies for dealing with disagreement between annotators. The label distribution and the inter-annotator agreement can provide a first indication of the difficulty of the task. A first impression of such data and annotation characteristics can be obtained by calculating common distributional statistics and inspecting the data statements if they are available [9, 10]. In annotation tasks that require annotators to generate text, such as caption generation, hypothesis generation for textual entailment, or question generation for question-answering tasks, the two aspects fall together because annotators might have stylistic or conceptual preferences that are not controlled by the annotation guidelines.

Knowing the characteristics of the data is so important because they influence the behavior that is learned by the model [11]. Machine learning models for classification are commonly regularized to find the simplest decision boundary between classes which still generalizes to validation data. This preference for simplicity is known as Occam's razor [12, 13]. Cognitive studies indicate that humans also adopt the simplicity principle when inferring patterns [14, 15]. However, if the observed data is not representative of the complexity of the problem, simple explanations will not generalize well. If the training data contains systematic gaps "that (unintentionally) allow simple decision rules to perform well on test data" [16], the resulting model might be a good fit for the data but not a cognitively plausible fit for the problem. In textual entailment tasks, for example, the entailment relation between a premise and a hypothesis can often be predicted by only looking at the premise and ignoring the hypothesis due to unintentional annotation preferences in the dataset generation [17]. This indicates that the learned patterns do not correspond to a cognitive notion of natural language understanding.

It is a challenging task to identify such systematic gaps because they are task-dependent and require knowledge about the annotation setup. Geva et al. [18] propose targeted analyses to identify annotator biases. They first show that the performance of the model improves if the id of the annotator is known. In a second step, they confirm that annotator identities can be predicted from the data and that the model overfits to known annotators and their lexical and stylistic choices. It is thus important that the annotators are representative of the scenario in which the model will be deployed to ensure generalizability. The undesirable consequences of biased datasets are further discussed in Sect. 5.4.

4.1.1.2 Subpopulations of Instances

In order to explore input–output patterns in a more hypothesis-driven way, we can analyze the behavior of the model for subpopulations with common properties. It is important to precisely define the scope of such subpopulations and different perspectives can play a role [19]. Subpopulations can be determined based on linguistic characteristics that are expected to have an influence on the task. For example, we might want to analyze *instances with multiple adjectives* for sentiment detection or *instances with relative clauses* for parsing. As a machine learning model can only learn generalizations over instances that are seen in the training data, subpopulations can also be defined relative to characteristics of the dataset. For example, instead of determining long instances by setting a threshold (e.g., *longer than 30 tokens*), Goel et al. [19] recommend defining length relatively (e.g., *instances that are in the 95th percentile for length*). For analyzing the performance of the model, characteristics of the annotation labels which need to be predicted can also be used to define subpopulations. For instance, we might want to inspect *all instances of class X* or *all instances which contain token labels X and Y in the same phrase*. Ethayarajh et al. [20] use the term slicing to refer to the process of identifying subpopulations with similar difficulty for the model.

Grouping erroneously predicted instances into manually identified categories is an established technique for error analysis [21]. However, random spot-checking of errors re-enforces confirmation bias [22]. Wu et al. [23] therefore postulate that error prevalence should be assessed over the entire dataset and provide the Errudite tool to automatically select groups of instances (e.g., *all instances that contain a named entity*).

When analyzing behavioral patterns, our goal is to better understand the strength and weaknesses of a model. As a consequence, the analysis of subpopulations may be led by prior assumptions about the model. For example, when we propose a new neural mechanism that is supposedly more useful for processing complex subclauses, we might want to analyze *all instances containing center embeddings*. Neural architectures are known to strongly rely on lexical patterns [24]. Gururangan et al. [25] analyze lexical characteristics by calculating the point-wise mutual information for each input token with respect to the target classes. This model-driven information can then be used to define subpopulations to explore (e.g., analyze *instances containing multiple tokens with PMI above X but for conflicting classes*.

Subpopulations can be analyzed quantitatively by contrasting aggregated evaluation metrics either for subpopulation pairs (e.g., compare the accuracy for *instances with more than X tokens* to that *for instances with less than Y tokens*) or with respect to the whole population. If subpopulations are based on continuous variables, correlational analyses can also be used [26]. As a side effect, these analyses can also shed light on the idiosyncratic characteristics of the dataset and support the detection of outliers and systematic gaps.

The analysis of subpopulations helps in better understanding the inductive bias of the model. A model which consistently struggles with a subpopulation can be more reliable than a model that makes fewer errors but follows less predictable patterns. The performance of the model is often deemed to be cognitively plausible if it aligns with the researchers' expectations. However, the choice of subpopulations to analyze is clearly affected by the individual preferences of the researchers and therefore subject to experimenter bias and confirmation bias (see Sect. 5.4).

4.1.2 Considering Difficulty

An important aspect of cognitive plausibility is the complexity of the decision. Within a task, the instances will vary in difficulty. We assume that a model that fails to predict the outcome for difficult instances is more cognitively plausible than a model that fails on easy instances.

In humans, we can measure the complexity of instances by the inter-annotator agreement or by the cognitive load (see Chap. 3). We can also determine the linguistic complexity of the input based on theoretical assumptions, e.g., several diagnostic datasets are based on grammar handbooks (see Sect. 4.2.1.2).

4.1.2.1 Instance-Level Evaluation

Zhong et al. [27] propose to compare models on the instance level instead of focusing on increased average performance. They provide an example of two BERT models which differ only by 0.1% in accuracy but assign different labels to 8% of the instances, i.e., true and false predictions are distributed differently. When comparing multiple models, an instance accuracy of 0.6, for example, indicates that three out of five models assign the correct label. They speculate that instance accuracy across models can be an indicator of noise in the dataset (i.e., debatable or wrong annotations) but found that this is only the case for a small fraction of the instances. Khurana et al. [28] also compare model performance on the instance level by calculating the overlap ratio of the wrong predictions with respect to specific perturbation categories.

4.1.2.2 Instance Difficulty

Standard evaluation scenarios in NLP expect a single label for each instance and rigorous procedures ensure that disagreements in annotation are resolved systematically. However, disagreement might also be a valuable signal of uncertainty for difficult or debatable cases. Plank et al. [29] investigate the role of disagreement in the annotation of part-of-speech tags and find that disagreement mostly indicates uncertainty for linguistically debatable cases. Pavlick and Kwiatkowski [30] analyze disagreement in textual entailment and come to similar conclusions. They propose that "NLI evaluation should explicitly incentivize models to predict distributions over human judgments." Zhang and de Marneffe [31] use an artificial annotator model to distinguish between systematic inference items and disagreement items.

Rondeau and Hazen [22] suggest a reverse approach and distinguish between easy, medium, and hard questions based on model performance. In this case, the model provides cues for learning more about a task. A closer look at the questions that are hard for the model can reveal insights into unconscious human processing patterns that rely on information that is not available to the model. Similarly, Zhan et al. [32] propose to conduct difficulty-aware evaluation by assigning larger weights to instances that are difficult to predict by most models. In this scenario, difficulty is not determined by human abilities but by the shortcomings of the models. In our opinion, these two perspectives on difficulty should align for cognitively plausible models: model uncertainty should be higher for instances that are difficult for humans.

When we draw conclusions from the performance of the model on the instance level, we do not take into account that the model might be right for the wrong reasons. An instance might be difficult for the model but still be solved correctly due to an incidental alignment with a learned shortcut or by belonging to the majority class. Quantifying the uncertainty of the model can be helpful for assessing the reliability of input–output patterns.

4.1.2.3 Model Uncertainty

In models, the complexity of an instance is often measured as uncertainty of the model in making a prediction. For classification tasks, models usually output a probability distribution over the output labels. In neural models, this is obtained by calculating the softmax function over the output activations. The model prediction is determined by choosing the label with the maximum probability. Instead of only considering the label with the maximum probability, the difficulty of a decision can be approximated by analyzing the characteristics of the probability distribution to determine the confidence of the model [30, 33, 34]. Model confidence usually refers to the probability assigned to the predicted class. As neural networks tend to be "over-confident", especially for noisy inputs [35], it is important to analyze whether the model confidence is aligned with its accuracy. This test is called model calibration [36]. If a model is calibrated, the relative probability differences between classes could be even more informative. A flat probability distribution over labels indicates lower model confidence, which can be interpreted as higher item difficulty. This is comparable to measuring perplexity for language models [37]. However, Baan et al. [38] show that measuring calibration is not meaningful for challenging tasks that cause disagreement between annotators.

Atanasova et al. [33] propose an alternative method for determining model confidence based on class-specific patterns of token-level saliency, which indicates the contribution of an input element to the probability of an output label (see Sect. 6.1). They interpret the distance between the saliency scores for the predicted output class to the saliency scores of the other classes as a confidence indication based and expect that an input token that has a high saliency score for one class is expected to have a low saliency score for other classes if the model is confident. In turn, small saliency distances indicate that the model is uncertain.

Schwartz et al. [39] propose that distinguishing between simple and hard instances based on model confidence can be used to reduce computational costs for predictions. Simple instances can be predicted faster using an early "exit" from neural network calculations to reduce computational costs whereas the full complexity of the model is used for hard instances.

It is essential to combine these methods with data analysis techniques for identifying biases in the dataset. A model might have learned the heuristic that instances that start with the token "however" always belong to the same class and assign high confidence to the prediction for a difficult instance despite misinterpreting it. These lazy shortcuts can often be revealed by isolating linguistic phenomena and testing the generalizability of a model to new data. For example, Bastings et al. [40] create partially synthetic data with explicit lexical shortcuts to analyze the sensibility of the model and find that attribution methods are very inconsistent in finding these shortcuts.

4.1.2.4 Learning Curves

Another way to interpret the difficulty of instances is to measure how much effort is required for the model to get them right. The effort can be quantified with respect to the size of the training data, the number of training epochs, or the number of parameters. When we teach language to children or to L2 learners, we start with a limited vocabulary and simple syntactic structures and gradually increase the difficulty based on the competence of the learner [41]. The input to language models is not curated based on difficulty, but their learning curves indicate that simpler linguistic phenomena are acquired earlier in the process. Warstadt et al. [42] show that the performance on specific linguistic phenomena improves steadily with more data while the general perplexity of the language model decreases. However, the rate of improvement varies depending on the difficulty of the phenomenon. Local phenomena such as determiner-noun agreement are acquired quite quickly, whereas distinguishing acceptable contexts for negative polarity items requires considerably more training instances.

When diagnostic instances are optimized to isolate certain linguistic factors, the resulting sentences often sound less natural and are unlikely to be frequent in training corpora. For example, Warstadt et al. [43] analyze how much pretraining data is needed before a model prefers linguistic generalizations over surface biases. In order to clearly distinguish multiple linguistic phenomena, they construct sentences such as *"These men weren't hating that this person who sang tunes destroyed the vase"* which seem overly artificial. On the lower end of the naturalness scale, we find experiments with artificial languages. They make it possible to systematically answer hypotheses about processes in neural models without having to pay attention to the irregular subtleties of authentic language. White and Cotterell [44] analyze the sensitivity of language models to typological differences between languages. They develop a grammar that uses "switches" to systematically vary typological features such as word order by re-arranging constituents while keeping all other choices constant to reduce potential confounding factors. Galke et al. [45] analyze the learnability of an artificial language and find that more structured languages lead to better generalization and memorization capabilities in neural models which is in line with findings for humans [46]. Most stimuli generated for experiments with artificial languages do not follow a Zipfian lexical distribution, which is typical for almost all languages. As language models are very sensitive to frequency effects, findings with artificial non-Zipfian distributions are not cognitively plausible and might not generalize well to real languages [47–49].

Swayamdipta et al. [50] instead use unchanged training instances and analyze the behavior of the model over training epochs. They monitor the variability of the model's confidence for the true class and distinguish between instances that are "easy to learn", "hard to learn", or "ambiguous". They propose that easy instances are most relevant for model optimization, ambiguous instances contribute to the model's ability to generalize to out-of-distribution data, and "hard" instances often correspond to labeling errors or outliers. Categorizing the training data in this way could help in identifying a useful order for curriculum learning (see Sect. 4.3.2).

4.2   Testing Behavior

The most widely known test for determining if the behavior of a model is cognitively plausible is the Turing test [51]. A model is considered to pass this test if a human interacting with the model cannot distinguish whether the responses come from a computational model or a human. This setup is expensive because it requires human participants, and the results are difficult to compare because the investigative procedure by the human remains unconstrained and is thus determined by subjective preferences. To evaluate the cognitive plausibility of models more directly, testing datasets automate the comparison to expected human queries.

We loosely distinguish between testing general linguistic phenomena in pretrained language models (Sect. 4.2.1) and more task-specific robustness tests for finetuned models (Sect. 4.2.2). In practice, the described methods usually overlap because linguistic knowledge is vital for almost all NLP tasks and generalizability is a key desideratum for pretrained language models [52].

4.2.1   Testing Linguistic Phenomena

Humans are able to generalize their language processing skills to sentences they have never seen before by applying compositional principles. Recent studies indicate that language models rely more strongly on memorization effects and fail to generalize to out-of-domain data [24]. Keller [53] had already specified "broad coverage" and "robust to noise" as important desiderata for cognitively plausible models. Recent discussions on what constitutes natural language understanding highlight the importance of compositionality for generalization skills [54, 55]. The development of systematic tests that approximate the linguistic abilities of a model is an ongoing subfield of interpretability research.

4.2.1.1   Minimal Pairs
Linguistic phenomena are often described by the difference in a minimal pair. For example, number agreement between subject and verb can be isolated by comparing a sentence with correct agreement (*The waiter **places** the glass on the table*) to the same sentence with incorrect agreement (*The waiter ***place** the glass on the table*. If the sentence becomes more

complex, the phenomenon remains the same, but identifying the correct sentence requires higher cognitive effort due to competing candidates for the subject: *The average of estimates of the 10 economists polled **places**/***place** the dollar around 1.820 marks.*[2]

Minimal pairs have been used extensively in controlled psycholinguistic experiments that aim to analyze specific linguistic phenomena (see Sect. 3.1). For instance, expectation-based theories of language comprehension, in particular surprisal theory, go a long way in accounting for the behavior of word-by-word processing difficulty (e.g., Harm Brouwer et al. [57]). Such minimal pairs can also be used to test the sensitivity of language models to specific linguistic phenomena. Each instance of the minimal pair is scored by the model by calculating the sum of the token log-probabilities. If a model consistently assigns a higher probability to the correct instance for a large range of examples, it can be concluded that the model is sensitive to this phenomenon [58]. A carefully curated suite of minimal pairs grouped by linguistic categories is called a diagnostic dataset.

4.2.1.2 Diagnostic Datasets

The creation of minimal pairs for testing simple agreement phenomena can be automated by replacing tokens in example sentences crawled from newspaper corpora or other curated resources. More complex linguistic phenomena such as negative polarity or reflexive anaphora require more sophisticated filtering of preprocessed data [59]. An even higher level of control can be achieved by automated generation procedures.

Marvin and Linzen [60] generate English examples using templates based on non-recursive context-free grammars. Warstadt et al. [42] follow a similar template-based approach with a larger vocabulary of 3,000 words to generate the BLiMP dataset. Their selection of linguistic phenomena is determined based on human curricula in syntax textbooks and covers a range of 12 phenomena from English morphology, syntax, and semantics. In a similar vein, Gauthier et al. [61] provide a tool for the creation of test suites of minimal pairs. Gulordava et al. [62] intentionally include nonce sentences to isolate the syntactic phenomenon and abstract from potential lexical, semantic, and frequency-based confounds. Such out-of-distribution stimuli can be problematic as they are clear outliers compared to the instances seen during training. The model is thus not able to represent them accurately, which might lead to erratic behavior.

Examining out-of-distribution behavior is useful for estimating the robustness of a language model in application scenarios, but the results can be misleading for generative models. Vamvas and Sennrich [63] show that distributional discrepancy between the diagnostic instances and the typical generations by the model obscure the interpretation. In their examples, both instances of the minimal pair were highly unlikely to be generated by the model, making the probability difference between the instances irrelevant for interpreting model behavior.

---

[2] The example is taken from [56] but we replaced the original main verb *puts* with *places* to avoid ambiguity with the past tense form.

The diagnostic datasets mostly target syntactic phenomena and focus on generating a large range of examples. Targeting semantic phenomena requires more manual control because the identification of selectional preferences and semantic roles involves finegrained lexical and world knowledge. Ettinger [34] uses stimuli that were manually designed by psycholinguistic experts to test the sensitivity of the BERT language model to semantic categories, role reversal, and negation. They compare the model's sensitivity to human cloze probabilities and to N400 responses in EEG (see also Sect. 6.2.2.2).

4.2.1.3 Winograd Schemas

A specific example of a diagnostic test of semantic capabilities are Winograd schemas, which were introduced by Levesque et al. [64]. A Winograd schema is a minimal sentence pair that contains a highly ambiguous pronoun and differs only in one or two tokens. Due to the lexical difference, the pronoun is resolved differently in the two sentences. In the example below from the WSC285 dataset [65], the difference between the two sentences is the adjective *large/small*.

1. *The trophy doesn't fit into the brown suitcase because it is too* **large**.
2. *The trophy doesn't fit into the brown suitcase because it is too* **small**.

In the first sentence, humans would resolve the pronoun *it* to refer to the trophy, whereas in the second sentence it refers to the suitcase. Resolving the anaphor requires commonsense knowledge about spatial relations. Winograd schemas can be used directly as diagnostic datasets, for example, to detect gender bias in language models [66]. It has become more common to transform the sentence pairs into a textual entailment task such that the premise corresponds to the full sentence and the pronoun is replaced with one of the candidate nouns for the hypothesis. The example above would be transformed into the following presentation:

> Premise: *The trophy doesn't fit into the brown suitcase because it is too* **large**.
> Hypothesis 1: *The trophy is too large.*
> Relation 2: *Entailment*
>
> Hypothesis 2: *The suitcase is too large.*
> Relation: *No Entailment*

For the version of the sentence with the adjective *small*, the entailment relations are reversed. Winograd schemas are challenging to create because Levesque et al. [64] defined a set of constraints to avoid that the schemas can be resolved by frequency effects and selectional preferences. In the example above, both *trophy* and *suitcase* should be more or less equally likely to occur with the adjective *large*. Approaches that automate the process therefore often yield sentences that violate at least one of the constraints or are difficult to understand [67]. Generally, psycholinguistically motivated diagnostic datasets that test more advanced skills of natural language understanding are too small to systematically finetune models and can only be used in tuning-free prompting scenarios. Linzen [68] proposes that such diagnostic benchmarks should only be applied for testing to avoid overfitting the model to unintended statistical regularities underlying the actual task. He claims that models should be favored that "like humans, are able to perform tasks with minimal instruction."

4.2.1.4 Test Suites

In order to test models on diverse aspects of natural language understanding, the GLUE benchmark suite has been developed [69]. GLUE provides a standardized environment for several diagnostic tasks including Winograd schemas. SuperGLUE has been proposed as a more challenging complementary benchmark that focuses on more finegrained linguistic phenomena [70]. All of the diagnostic datasets discussed here target the English language. Both the GLUE and SuperGLUE benchmarks have been translated into multiple other languages, see Kocijan et al. [67] for an overview. Some cases needed to be excluded or adjusted because the translation resolves the ambiguity in the original sentence due to grammatical gender cues. This indicates that cognitively plausible behavior is strongly influenced by the language under study. Diagnostic datasets that target specific phenomena of non-English languages are still scarce. Ruder [71] proposes that future benchmarks should comprise multiple tasks in multiple languages and test the generalization abilities of models in a finegrained and ideally adaptive way. The Beyond the Imitation Game (BIG-bench) project targets this goal in a collaborative effort and already collected more than 200 test tasks in multiple languages [72].

4.2.2 Robustness and Generalizability

While the diagnostic datasets aim at examining whether pretrained language models have learned general linguistic capabilities, more task-specific methods can test the robustness of a finetuned model.

4.2.2.1 Occlusion Tests

An exploratory approach for identifying decision flips are occlusion methods which quantify the relevance of each input token to the model's prediction by iteratively removing them. When the contributions of all possible combinations of input elements are calculated, the resulting scores are often considered to be Shapley values, a concept originating in game theory [73, 74].

Occlusions methods have been criticized because they create ungrammatical instances that are out of the training distribution and therefore cannot provide reliable information about the performance of the model for in-distribution instances (see also Sect. 4.2.1.2).

Furthermore, occlusion analyses cannot capture patterns related to tokens absent from the input [75], and can only approximate compositional effects [76].

Jumelet et al. [77] show how a Shapley-based technique called contextualizeds decomposition can be applied in a more hypothesis-driven way. They distinguish between relevant and irrelevant tokens for analyzing number agreement and decompose the model's activation values into these two groups. In order to test such hypotheses more explicitly, linguistically meaningful perturbations can be applied to the input.

4.2.2.2 Perturbations and Decision Flips

Perturbations are systematic modifications to existing test instances with a clear cognitively grounded expectation on how the perturbation affects the label. Ribeiro et al. [78] distinguish between invariance tests and directionality tests and provide the Checklist tool to systematically generate them. In order to understand the generalization capabilities of a model, perturbations that lead to a cognitively unexpected decision flip are particularly interesting [79].

Invariance is tested by perturbing the input in a way that preserves the label. For example, the introduction of typos or the variation of named entities is generally not expected to change the sentiment of a sentence. A subset of invariance tests are so-called adversarial examples [80]. They are intentionally created in a way to "fool" the model based on prior assumptions about its behavior. Adversarial examples simulate users that try to exploit gaps in the model, for example, by adding meaningless jargon to an essay submitted to an automatic scoring system [81] or by replacing characters in swearwords ("f*ck") to get around hatespeech detection [82].

For directional tests, the expectation indicates that the perturbation may only lead to a label change in a specified direction. For example, adding phrases that are known to be associated with negative sentiment should not trigger a label change toward positive sentiment. The predicted sentiment should stay the same or become more negative after the perturbation. The precise definition of expectations for the effect of the perturbations makes it possible to quantitatively evaluate the robustness of the model. Perturbations do not require additional ground truth labels from humans but rather test cognitively plausible relationships between the original instance and the perturbed instance. A subset of directional tests are counterfactual examples that are expected to lead to a label change in the opposite direction, for example, by adding negation.

If we manipulate instances automatically, we often accidentally create examples that are ungrammatical or cognitively implausible. Jia et al. [83] test invariance by substituting tokens with near-synonyms according to static list, for example, by changing *very* to *supremely*. Such changes do not necessarily maintain semantic coherence and can lead to unnatural utterances. The observed model behavior for such outliers in the input distribution cannot be robustly interpreted as a decision pattern because it only shows how the model deals with unexpected inputs [35, 84].

4.2.2.3 Influence of the Training Data

Systematic exploration of decision flips can help to identify the decision boundaries of the model. Influence functions indicate more directly how the presence of an instance in the training data influences the model's decision for a test instance [85]. They approximate leave-one-out testing and track the parameter change in the model when perturbing the loss for a particular instance. High influence scores indicate that the presence of a training instance increases the probability of the model's prediction. As the calculation of the influence scores is extremely time-intensive, more efficient approximations have been developed, see Sun et al. [86] for an overview.

If we notice that the model relies on a latent statistic property of the data which is not relevant for the task, we can attempt to develop a contrast set to reduce this systematic gap in the training data as proposed by Gardner et al. [16]. For example, they find that in the multiple-choice question-answering dataset DROP [87], a good approach to answer questions that start with *"How many...?"* is to always pick the second answer and that questions which ask about the ordering of events typically follow the order of the paragraphs. We can test if a model exploits these simple strategies by manually creating a task-specific contrast set in which the order of answers and events are altered. Gardner et al. [16] report significant performance drops in state-of-the-art models on a collection of ten contrast sets for diverse tasks, while the human performance remained consistent for the original dataset *and* the contrast set.

Perturbations and generalizability tests go beyond the explorative analysis of input–output patterns by defining clear testing criteria and formulating cognitively grounded expectations for the model behavior. If the model fails a test systematically, we have identified cognitively unexpected behavior. We can then try to isolate the root of this behavior by comparing its variance across different modeling decisions (see Chap. 2).

4.3   Towards Cognitively Plausible Behavior

We have described methods for analyzing and testing the cognitive plausibility of behavioral patterns of language models. The hypothesis space of behavioral analyses still remains largely underexplored primarily due to the prevailing reliance on aggregated quantitative evaluations in NLP. A common criticism of neural models is the lack of interpretability associated with them. In recent years, exciting progress has been made in developing interpretability methods that allow us to peek inside the mechanisms of neural computation. We delve deeper into these methods in the following chapters but we think that significant progress towards cognitively more plausible models can also be achieved when remaining purely on the behavioral level. Three promising directions are methods for finegrained evaluation, curriculum learning, and a wider multilingual perspective.

4.3.1 Finegrained Evaluation

Development in natural language processing is driven by quantitative evaluation on gold standard datasets. In order to incentivize the development of cognitively more plausible models, we need to integrate cognitive criteria into benchmark evaluation. Test suites that combine diverse diagnostic tests make it possible to compare how modeling decisions affect linguistic generalization abilities [70]. Ruder and Sil [88] propose to cover diverse aspects of language understanding in the evaluation to assess the generalizability of the model on a range of tasks. Unfortunately, the performance on such benchmarks is often aggregated into a single score which hampers a deeper analysis. Therefore, we propose that the comparison of two models for a given evaluation task should also be quantified on the instance level by calculating instance accuracy [27] and overlap ratios [28]. These metrics should be cross-referenced with measures for instance difficulty such as annotator disagreement or cognitive load, and differences in model uncertainty such as distributions of confidence and saliency scores. When it comes to annotated data, raw annotations should always be provided to assess disagreement and outliers. Hao et al. [89] point out that predictability norm correlations with human annotations are a cognitively more plausible evaluation measure for language models than text-driven perplexity. Pavlick and Kwiatkowski [30] propose that models should predict distributions over human judgments and Simpson and Gurevych [90] use Bayesian preference learning over crowd annotations to account for individual preferences instead of resorting to a skewed majority vote. Their model can also account for sequential dependencies [91]. To approximate Bayesian reasoning with neural models, Monte-Carlo dropout has been proposed as a way to model uncertainty [92]. For this method, dropout layers are not only applied during training but also during testing, leading to variation in the prediction. Mean and variance of the distribution over the output labels are then obtained by averaging over multiple forward passes through the network with the same input data but experiments on visual datasets indicate that the uncertainty estimate is strongly influenced by experimental parameters such as the dropout rate [93]. Recent experiments on applying Monte-Carlo dropout to natural language processing problems [94–96] can be the foundation for a new experimental paradigm that moves away from targeting the dogmatic one-dimensional gold standard truth towards cognitively more plausible models that account for uncertainty and disagreement.

4.3.2 Curriculum Learning

In neural network training, the examples for each batch are commonly chosen at random. However, when humans teach, they start with easy or typical examples first, before introducing more complex or ambiguous examples. Studies in cognitive science have confirmed that humans learn much better when training examples are not randomly presented, but organized into a sequence of easy to hard examples [97, 98]. Schluter and Varab [99] show that the order of training instances also has an effect on computational models: simply shuffling the order of training instances can lead to significant performance drops for a model optimized for a natural language inference task. Their approach can be an important sanity check for the robustness of quantitative results.

Curriculum learning has been proposed as a learning paradigm that applies knowledge-driven batch sampling when training neural networks. To implement a cognitively more plausible curriculum, two challenges need to be tackled: (1) sorting the examples in the training data by difficulty, and (2) finding a good pacing function to compute batches with an increasing level of difficulty over time [100]. When these two challenges are met, curriculum learning can lead to performance improvements without additional computational costs [101] and improve the cognitive plausibility of the learning process. We have discussed data-driven methods to rank the samples from easy to difficult. If available, cognitive data, such as eye-tracking fixations or reading times, can provide additional insights into the difficulty of instances. Combining these signals with cognitive and educational theories of learning can help to categorize difficulty classes for instances and to find the right pacing function for introducing more difficult data [102].

The effect of curriculum learning on language model pretraining has been explored by Campos [103] who approximate instance difficulty using linguistic properties in line with linguistic theories of sentence complexity. They compare surface measures such as sentence length with more advanced semantic and syntactic complexity measures (e.g., sentence entropy, dependency depth, diversity in part-of-speech tags) and evaluate transfer performance on the GLUE benchmark [69]. They report that the difference between curriculum-informed batching and random batching was minimal when evaluating general perplexity and task-based averages. Interestingly, they find that their competence-based curriculum learning leads to improved performance on transfer tasks when the pretraining corpus is small. This indicates that it could be a cognitively more plausible approach with complementary strengths compared to the common data-hungry methods. In another direction, Soviany et al. [101] propose to link the curriculum criteria to vocabulary size because when humans learn a language, we also start with a limited vocabulary that progressively expands [104]. Willems et al. [105] propose an online criterium for curriculum learning with the idea that suitable next instances are the ones that are "learnable but not learned yet" and determine instance ordering based one the mastering rate of the model. In a similar vain, Lalor and Yu [106] propose to select instances by calculating an ability estimation of the model.

On a more task-specific level, syntactic approximations of text complexity have been used to inform curriculum learning for machine translation [107] and natural language generation [108]. Sachan and Xing [109] experiment with model-driven uncertainty metrics as a difficulty criterion for curriculum learning for question answering. As an additional constraint for cognitively plausible curriculum learning, they introduce the concept of instance diversity. They propose that instances should not only be ordered in increasing order of difficulty but that the dissimilarity between training instances should also be factored in to represent the variety of a task.

4.3.3 Multilingual Perspective

In this chapter, we focused on data analysis and evaluation on the instance level and linguistically informed diagnostic testing to identify model behavior that goes against cognitive expectations. As NLP researchers, we are often tempted to consider ourselves as the benchmark for cognitively plausible linguistic behavior. We are human, we can communicate and it is our job to think about language. Unfortunately, if we use our intuition, we often remain ignorant of linguistic characteristics that are not relevant to the languages we speak.

The need for a wider multilingual perspective in cognitive modeling had already been pointed out by Keller [53]. In recent years, multilingual models which are trained on more than one hundred languages have been developed and yield impressive performances for a wide range of tasks [110] despite being based on linguistically and cognitively naïve assumptions.

Language modeling architectures have been primarily developed and optimized for English, but Joshi et al. [111] find that 48% of all typological feature categories in the *World Atlas of Language Structures* exist only in low-resource languages. As crosslingual models treat all languages uniformly, it does not come as a surprise that their performance on low-resource languages still cannot compete with the results for high-resource languages.

Cabello Piqueras and Søgaard [112] aim at quantifying this phenomenon of cross-lingual unfairness. They developed a cloze dataset and find that the quality of sentence completion of the most popular multilingual language models varies even for closely related languages. An important aspect of language modeling quality is the choice of the unit (see Chap. 2). Liang et al. [113] show that using a shared subword vocabulary for all languages leads to over-tokenization for low-resource languages. They propose to increase the vocabulary to better capture cross-lingual diversity and to de-emphasize token sharing between languages with little lexical overlap.

Malik-Moraleda et al. [114] explore cross-linguistic differences in the neural implementation of language and find that variability observed across languages is similar to or lower than the inter-individual variability among speakers of the same language. Their sample is quite small with only two speakers per language but it highlights the need to better account for individual differences in cognitive analyses in order to better capture the underlying similarities and differences in the processing patterns.

We think that the combination of revisiting computational modeling choices in systematic experiments and integrating psycholinguistics and typological insights is crucial to derive a better understanding of the shared cognitive and neural bases of language processing [115]. This will help us disentangle modeling choices that are plausible for processing English from cognitively plausible modeling choices for processing language.

4.4   Ethical Aspects

When we treat human behavior as the gold standard for a model, we need to keep in mind that human cognition has flaws. We often take lazy shortcuts and rely on prototype knowledge and rules of thumb to take decisions [116]. This behavior is reflected in the data. van Miltenburg [117] shows that a dataset of crowdsourced image descriptions is not purely descriptive. The chosen focus of the descriptions provides useful information about the relative importance of objects and events. However, the descriptions also contain subjective inferences that cannot be drawn directly from the scene, for example, about relationships between persons or about their status. These unwarranted inferences are often characterized by stereotypes, e.g. when annotators assume that a male person is the boss of a female person although the image does not provide this information.

If we directly use behavioral datasets to train our models, we re-enforce societal biases about gender and race in the applications (see also Sect. 2.3). Bender et al. [118] propose to use the philosophical distinction between descriptive and normative ethics to characterize the behavior of our models. Descriptive ethics refers to modeling human behavior as we observe it in the world, normative ethics refers to modeling ideal behavior that we are striving towards. Finding the right gradation between the two strategies is a political question but it also affects very concrete choices in natural language applications. For instance, when thinking about applications that target children, it is evident that we should filter out harmful information.

When we discuss the cognitive plausibility of a model, one can argue that a model is more cognitively plausible if it also reproduces undesirable human patterns. We followed this line of thinking with respect to the difficulty of items: it is more acceptable if a model fails on difficult instances than on easy instances because this behavior can be more easily anticipated by human users. Even more desirable than a model that fails in a cognitively plausible way is a model that does not fail at all but reports lower confidence for difficult decisions. This can also be our directive toward ethical behavior. We do not want to reproduce ethically flawed decisions but provide a more transparent account of the societal biases that can be involved. If we can trace back which factors influence the decisions in computational models, we might also learn more about human behavior.

The trade-off between normative and descriptive modeling needs to be evaluated with respect to individual applications. We need different language models in educational scenarios than when modeling the political sentiment of subgroups. In order to identify imbalances and biases in the data and potentially employ countermeasures during finetuning, it is imperative to facilitate access to the pretraining data of language models (see Sect. 2.3).

The choice of the pre-training data for language models illustrates the problem of measurement bias, also known as reporting bias [119]. The way we operationalize and measure cognitive concepts (e.g., understanding) affects what we will observe. For example, the perception of gradual criteria depends on the definition of the upper and lower bound. When we provide examples of perceived difficulty or similarity in annotation guidelines, they serve as anchors to gauge the scale of the variable. This can lead to more homogeneous annotations but the choice of the reference values also biases the annotators' judgment (see also Sect. 5.1.1). Another example of measurement bias for behavioral analysis is the definition of boundaries for subgroups. The criteria for the inclusion of borderline cases can have a big influence on our conclusions.

Behavioral analyses treat the model as a black box and only focus on input–output patterns. This can lead to wrong conclusions when a model is "right for the wrong reasons", e.g., by applying simplified heuristics that fortuitously lead to the correct result [120]. Guest and Martin [121] point out that simple correlations between a model and behavioral or brain imaging data may not lead to the conclusion that the model is human-like. They argue that a good fit between a model and cognitive signals is a necessary condition for assuming cognitive plausibility, but not a sufficient one because it lacks theoretical and empirical support. We have sketched out some diagnostic techniques for narrowing down observations of behavioral patterns. If we want to come closer to fulfilling the "right to explanation"[3] for language modeling use cases, we need to dive deeper into the representational structure and the procedural strategies of the model.

References

1. Nick Chater, Joshua B Tenenbaum, and Alan Yuille. Probabilistic models of cognition: Conceptual foundations, 2006.
2. John B Watson. Psychology as the behaviorist views it. *Psychological review*, 20 (2): 158, 1913.
3. Yvette Graham, Christian Federmann, Maria Eskevich, and Barry Haddow. Assessing human-parity in machine translation on the segment level. In *Findings of the Association for Computational Linguistics: EMNLP 2020*, pages 4199–4207, Online, November 2020. Association for Computational Linguistics. https://doi.org/10.18653/v1/2020.findings-emnlp.375. https://aclanthology.org/2020.findings-emnlp.375.
4. Kaiyang Zhou, Ziwei Liu, Yu Qiao, Tao Xiang, and Chen Change Loy. Domain generalization: A survey. *IEEE Transactions on Pattern Analysis and Machine Intelligence*, 2022b.
5. Alan Ramponi and Barbara Plank. Neural unsupervised domain adaptation in NLP—A survey. In *Proceedings of the 28th International Conference on Computational Linguistics*, pages 6838–6855, Barcelona, Spain (Online), December 2020. International Committee on Computational Linguistics. https://doi.org/10.18653/v1/2020.coling-main.603. https://aclanthology.org/2020.coling-main.603.
6. Nan Du, Yanping Huang, Andrew M Dai, Simon Tong, Dmitry Lepikhin, Yuanzhong Xu, Maxim Krikun, Yanqi Zhou, Adams Wei Yu, Orhan Firat, et al. Glam: Efficient scaling of language models with mixture-of-experts. In *International Conference on Machine Learning*, pages 5547–5569. PMLR, 2022.
7. Aakanksha Chowdhery, Sharan Narang, Jacob Devlin, Maarten Bosma, Gaurav Mishra, Adam Roberts, Paul Barham, Hyung Won Chung, Charles Sutton, Sebastian Gehrmann, et al. Palm: Scaling language modeling with pathways. *arXiv preprint* arXiv:2204.02311, 2022.

---

[3] For an interesting discussion about the fuzzy definition of the "right to explanation", see Chapter 12.3 of Søgaard [122].

8. Dirk Hovy and Shannon L. Spruit. The social impact of natural language processing. In *Proceedings of the 54th Annual Meeting of the Association for Computational Linguistics (Volume 2: Short Papers)*, pages 591–598, Berlin, Germany, August 2016. Association for Computational Linguistics. https://doi.org/10.18653/v1/P16-2096. https://aclanthology.org/P16-2096.

9. Emily M. Bender and Batya Friedman. Data statements for natural language processing: Toward mitigating system bias and enabling better science. *Transactions of the Association for Computational Linguistics*, 6:587–604, 2018. https://doi.org/10.1162/tacl_a_00041. https://aclanthology.org/Q18-1041.

10. Timnit Gebru, Jamie Morgenstern, Briana Vecchione, Jennifer Wortman Vaughan, Hanna Wallach, Hal Daumé Iii, and Kate Crawford. Datasheets for datasets. *Communications of the ACM*, 64 (12): 86–92, 2021.

11. Amandalynne Paullada, Inioluwa Deborah Raji, Emily M Bender, Emily Denton, and Alex Hanna. Data and its (dis) contents: A survey of dataset development and use in machine learning research. *Patterns*, 2 (11): 100336, 2021.

12. Alan Baker. Simplicity. In Edward N. Zalta, editor, *The Stanford Encyclopedia of Philosophy*. Metaphysics Research Lab, Stanford University, Summer 2022 edition, 2022.

13. Anselm Blumer, Andrzej Ehrenfeucht, David Haussler, and Manfred K Warmuth. Occam's razor. *Information processing letters*, 24 (6): 377–380, 1987.

14. Jacob Feldman. The simplicity principle in human concept learning. *Current Directions in Psychological Science*, 12(6):227–232, 2003. https://doi.org/10.1046/j.0963-7214.2003.01267.x. https://doi.org/10.1046/j.0963-7214.2003.01267.x.

15. Nick Chater and Paul Vit?nyi. Simplicity: a unifying principle in cognitive science? *Trends in Cognitive Sciences*, 7 (1): 19–22, 2003. ISSN 1364-6613. https://doi.org/10.1016/S1364-6613(02)00005-0. URL https://www.sciencedirect.com/science/article/pii/S1364661302000050.

16. Matt Gardner, Yoav Artzi, Victoria Basmov, Jonathan Berant, Ben Bogin, Sihao Chen, Pradeep Dasigi, Dheeru Dua, Yanai Elazar, Ananth Gottumukkala, Nitish Gupta, Hannaneh Hajishirzi, Gabriel Ilharco, Daniel Khashabi, Kevin Lin, Jiangming Liu, Nelson F. Liu, Phoebe Mulcaire, Qiang Ning, Sameer Singh, Noah A. Smith, Sanjay Subramanian, Reut Tsarfaty, Eric Wallace, Ally Zhang, and Ben Zhou. Evaluating models' local decision boundaries via contrast sets. In *Findings of the Association for Computational Linguistics: EMNLP 2020*, pages 1307–1323, Online, November 2020. Association for Computational Linguistics. https://doi.org/10.18653/v1/2020.findings-emnlp.117. https://aclanthology.org/2020.findings-emnlp.117.

17. Adam Poliak, Jason Naradowsky, Aparajita Haldar, Rachel Rudinger, and Benjamin Van Durme. Hypothesis only baselines in natural language inference. In *Proceedings of the Seventh Joint Conference on Lexical and Computational Semantics*, pages 180–191, New Orleans, Louisiana, June 2018. Association for Computational Linguistics. https://doi.org/10.18653/v1/S18-2023. https://aclanthology.org/S18-2023.

18. Mor Geva, Yoav Goldberg, and Jonathan Berant. Are we modeling the task or the annotator? an investigation of annotator bias in natural language understanding datasets. In *Proceedings of the 2019 Conference on Empirical Methods in Natural Language Processing and the 9th International Joint Conference on Natural Language Processing (EMNLP-IJCNLP)*, pages 1161–1166, Hong Kong, China, November 2019. Association for Computational Linguistics. https://doi.org/10.18653/v1/D19-1107. https://aclanthology.org/D19-1107.

19. Karan Goel, Nazneen Fatema Rajani, Jesse Vig, Zachary Taschdjian, Mohit Bansal, and Christopher Ré. Robustness gym: Unifying the NLP evaluation landscape. In *Proceedings of the 2021 Conference of the North American Chapter of the Association for Computational Linguistics: Human Language Technologies: Demonstrations*, pages 42–55, Online, June 2021. Associa-

tion for Computational Linguistics. https://doi.org/10.18653/v1/2021.naacl-demos.6. https://aclanthology.org/2021.naacl-demos.6.

20. Kawin Ethayarajh, Yejin Choi, and Swabha Swayamdipta. Understanding dataset difficulty with $\mathcal{V}$-usable information. In Kamalika Chaudhuri, Stefanie Jegelka, Le Song, Csaba Szepesvari, Gang Niu, and Sivan Sabato, editors, *Proceedings of the 39th International Conference on Machine Learning*, volume 162 of *Proceedings of Machine Learning Research*, pages 5988–6008. PMLR, 17–23 Jul 2022. https://proceedings.mlr.press/v162/ethayarajh22a.html.

21. Jonathan K. Kummerfeld, David Hall, James R. Curran, and Dan Klein. Parser showdown at the Wall Street corral: An empirical investigation of error types in parser output. In *Proceedings of the 2012 Joint Conference on Empirical Methods in Natural Language Processing and Computational Natural Language Learning*, pages 1048–1059, Jeju Island, Korea, July 2012. Association for Computational Linguistics. https://aclanthology.org/D12-1096.

22. Marc-Antoine Rondeau and T. J. Hazen. Systematic error analysis of the Stanford question answering dataset. In *Proceedings of the Workshop on Machine Reading for Question Answering*, pages 12–20, Melbourne, Australia, July 2018. Association for Computational Linguistics. https://doi.org/10.18653/v1/W18-2602. https://aclanthology.org/W18-2602.

23. Tongshuang Wu, Marco Tulio Ribeiro, Jeffrey Heer, and Daniel Weld. Errudite: Scalable, reproducible, and testable error analysis. In *Proceedings of the 57th Annual Meeting of the Association for Computational Linguistics*, pages 747–763, Florence, Italy, July 2019. Association for Computational Linguistics. https://doi.org/10.18653/v1/P19-1073. https://aclanthology.org/P19-1073.

24. Aparna Elangovan, Jiayuan He, and Karin Verspoor. Memorization vs. generalization : Quantifying data leakage in NLP performance evaluation. In *Proceedings of the 16th Conference of the European Chapter of the Association for Computational Linguistics: Main Volume*, pages 1325–1335, Online, April 2021. Association for Computational Linguistics. https://doi.org/10.18653/v1/2021.eacl-main.113. https://aclanthology.org/2021.eacl-main.113.

25. Suchin Gururangan, Swabha Swayamdipta, Omer Levy, Roy Schwartz, Samuel Bowman, and Noah A. Smith. Annotation artifacts in natural language inference data. In *Proceedings of the 2018 Conference of the North American Chapter of the Association for Computational Linguistics: Human Language Technologies, Volume 2 (Short Papers)*, pages 107–112, New Orleans, Louisiana, June 2018. Association for Computational Linguistics. https://doi.org/10.18653/v1/N18-2017. https://aclanthology.org/N18-2017.

26. Jonathan Kamp, Lisa Beinborn, and Antske Fokkens. Perturbations and subpopulations for testing robustness in token-based argument unit recognition. In *Proceedings of the 9th Workshop on Argument Mining*, pages 62–73, Online and in Gyeongju, Republic of Korea, October 2022. International Conference on Computational Linguistics. https://aclanthology.org/2022.argmining-1.5.

27. Ruiqi Zhong, Dhruba Ghosh, Dan Klein, and Jacob Steinhardt. Are larger pretrained language models uniformly better? comparing performance at the instance level. In *Findings of the Association for Computational Linguistics: ACL-IJCNLP 2021*, pages 3813–3827, Online, August 2021. Association for Computational Linguistics. https://doi.org/10.18653/v1/2021.findings-acl.334. https://aclanthology.org/2021.findings-acl.334.

28. Urja Khurana, Eric Nalisnick, and Antske Fokkens. How emotionally stable is ALBERT? testing robustness with stochastic weight averaging on a sentiment analysis task. In *Proceedings of the 2nd Workshop on Evaluation and Comparison of NLP Systems*, pages 16–31, Punta Cana, Dominican Republic, November 2021. Association for Computational Linguistics. https://doi.org/10.18653/v1/2021.eval4nlp-1.3. https://aclanthology.org/2021.eval4nlp-1.3.

29. Barbara Plank, Dirk Hovy, and Anders Søgaard. Linguistically debatable or just plain wrong? In *Proceedings of the 52nd Annual Meeting of the Association for Computational Linguis-*

tics (Volume 2: Short Papers), pages 507–511, Baltimore, Maryland, June 2014. Association for Computational Linguistics. https://doi.org/10.3115/v1/P14-2083. https://aclanthology.org/P14-2083.

30. Ellie Pavlick and Tom Kwiatkowski. Inherent disagreements in human textual inferences. *Transactions of the Association for Computational Linguistics*, 7:677–694, 2019. https://doi.org/10.1162/tacl_a_00293. https://aclanthology.org/Q19-1043.

31. Xinliang Frederick Zhang and Marie-Catherine de Marneffe. Identifying inherent disagreement in natural language inference. In *Proceedings of the 2021 Conference of the North American Chapter of the Association for Computational Linguistics: Human Language Technologies*, pages 4908–4915, Online, June 2021. Association for Computational Linguistics. https://doi.org/10.18653/v1/2021.naacl-main.390. https://aclanthology.org/2021.naacl-main.390.

32. Runzhe Zhan, Xuebo Liu, Derek F. Wong, and Lidia S. Chao. Difficulty-aware machine translation evaluation. In *Proceedings of the 59th Annual Meeting of the Association for Computational Linguistics and the 11th International Joint Conference on Natural Language Processing (Volume 2: Short Papers)*, pages 26–32, Online, August 2021. Association for Computational Linguistics. https://doi.org/10.18653/v1/2021.acl-short.5. https://aclanthology.org/2021.acl-short.5.

33. Pepa Atanasova, Jakob Grue Simonsen, Christina Lioma, and Isabelle Augenstein. A diagnostic study of explainability techniques for text classification. In *Proceedings of the 2020 Conference on Empirical Methods in Natural Language Processing (EMNLP)*, pages 3256–3274, Online, November 2020. Association for Computational Linguistics. https://doi.org/10.18653/v1/2020.emnlp-main.263. https://aclanthology.org/2020.emnlp-main.263.

34. Allyson Ettinger. What BERT is not: Lessons from a new suite of psycholinguistic diagnostics for language models. *Transactions of the Association for Computational Linguistics*, 8:34–48, 2020. https://doi.org/10.1162/tacl_a_00298. https://aclanthology.org/2020.tacl-1.3.

35. Shi Feng, Eric Wallace, Alvin Grissom II, Mohit Iyyer, Pedro Rodriguez, and Jordan Boyd-Graber. Pathologies of neural models make interpretations difficult. In *Proceedings of the 2018 Conference on Empirical Methods in Natural Language Processing*, pages 3719–3728, Brussels, Belgium, October-November 2018. Association for Computational Linguistics. https://doi.org/10.18653/v1/D18-1407. https://aclanthology.org/D18-1407.

36. Shrey Desai and Greg Durrett. Calibration of pre-trained transformers. In *Proceedings of the 2020 Conference on Empirical Methods in Natural Language Processing (EMNLP)*, pages 295–302, Online, November 2020. Association for Computational Linguistics. https://doi.org/10.18653/v1/2020.emnlp-main.21. https://aclanthology.org/2020.emnlp-main.21.

37. Jennifer Hu, Jon Gauthier, Peng Qian, Ethan Wilcox, and Roger Levy. A systematic assessment of syntactic generalization in neural language models. In *Proceedings of the 58th Annual Meeting of the Association for Computational Linguistics*, pages 1725–1744, Online, July 2020. Association for Computational Linguistics. https://doi.org/10.18653/v1/2020.acl-main.158. https://aclanthology.org/2020.acl-main.158.

38. Joris Baan, Wilker Aziz, Barbara Plank, and Raquel Fernandez. Stop measuring calibration when humans disagree. In *Proceedings of the 2022 Conference on Empirical Methods in Natural Language Processing*, pages 1892–1915, Abu Dhabi, United Arab Emirates, December 2022. Association for Computational Linguistics. https://aclanthology.org/2022.emnlp-main.124.

39. Roy Schwartz, Gabriel Stanovsky, Swabha Swayamdipta, Jesse Dodge, and Noah A. Smith. The right tool for the job: Matching model and instance complexities. In *Proceedings of the 58th Annual Meeting of the Association for Computational Linguistics*, pages 6640–6651, Online, July 2020. Association for Computational Linguistics. https://doi.org/10.18653/v1/2020.acl-main.593. https://aclanthology.org/2020.acl-main.593.

40. Jasmijn Bastings, Sebastian Ebert, Polina Zablotskaia, Anders Sandholm, and Katja Filippova. "will you find these shortcuts?" a protocol for evaluating the faithfulness of input salience methods for text classification. In *Proceedings of the 2022 Conference on Empirical Methods in Natural Language Processing*, pages 976–991, Abu Dhabi, United Arab Emirates, December 2022. Association for Computational Linguistics. https://aclanthology.org/2022.emnlp-main.64.

41. Jack C Richards. *Curriculum development in language teaching*. Cambridge University Press, 2001.

42. Alex Warstadt, Alicia Parrish, Haokun Liu, Anhad Mohananey, Wei Peng, Sheng-Fu Wang, and Samuel R. Bowman. BLiMP: The benchmark of linguistic minimal pairs for English. *Transactions of the Association for Computational Linguistics*, 8: 377–392, 2020a. https://doi.org/10.1162/tacl_a_00321. https://aclanthology.org/2020.tacl-1.25.

43. Alex Warstadt, Yian Zhang, Xiaocheng Li, Haokun Liu, and Samuel R. Bowman. Learning which features matter: RoBERTa acquires a preference for linguistic generalizations (eventually). In *Proceedings of the 2020 Conference on Empirical Methods in Natural Language Processing (EMNLP)*, pages 217–235, Online, November 2020b. Association for Computational Linguistics. https://doi.org/10.18653/v1/2020.emnlp-main.16. https://aclanthology.org/2020.emnlp-main.16.

44. Jennifer C. White and Ryan Cotterell. Examining the inductive bias of neural language models with artificial languages. In *Proceedings of the 59th Annual Meeting of the Association for Computational Linguistics and the 11th International Joint Conference on Natural Language Processing (Volume 1: Long Papers)*, pages 454–463, Online, August 2021. Association for Computational Linguistics. https://doi.org/10.18653/v1/2021.acl-long.38. https://aclanthology.org/2021.acl-long.38.

45. Lukas Galke, Yoav Ram, and Limor Raviv. What makes a language easy to deep-learn? *arXiv preprint* arXiv:2302.12239, 2023.

46. Limor Raviv, Marianne de Heer Kloots, and Antje Meyer. What makes a language easy to learn? a preregistered study on how systematic structure and community size affect language learnability. *Cognition*, 210: 104620, 2021. ISSN 0010-0277. https://doi.org/10.1016/j.cognition.2021.104620. URL https://www.sciencedirect.com/science/article/pii/S0010027721000391.

47. Andrew T Hendrickson and Amy Perfors. Cross-situational learning in a zipfian environment. *Cognition*, 189: 11–22, 2019.

48. Amir Shufaniya and Inbal Arnon. A cognitive bias for zipfian distributions? uniform distributions become more skewed via cultural transmission. *Journal of Language Evolution*, 7(1):59–80, 2022.

49. Ori Lavi-Rotbain and Inbal Arnon. The learnability consequences of zipfian distributions in language. *Cognition*, 223:105038, 2022.

50. Swabha Swayamdipta, Roy Schwartz, Nicholas Lourie, Yizhong Wang, Hannaneh Hajishirzi, Noah A. Smith, and Yejin Choi. Dataset cartography: Mapping and diagnosing datasets with training dynamics. In *Proceedings of the 2020 Conference on Empirical Methods in Natural Language Processing (EMNLP)*, pages 9275–9293, Online, November 2020. Association for Computational Linguistics. https://doi.org/10.18653/v1/2020.emnlp-main.746. https://aclanthology.org/2020.emnlp-main.746.

51. A.M. Turing. Computing machinery and intelligence. *Computing Machinery and Intelligence*, page 433–460, 1950. URL https://www.scopus.com/inward/record.uri?eid=2-s2.0-0011983060&partnerID=40&md5=b99ae3ebef56a66e44fe24b19073c0d8. Cited by: 191.

52. Dieuwke Hupkes, Mario Giulianelli, Verna Dankers, Mikel Artetxe, Yanai Elazar, Tiago Pimentel, Christos Christodoulopoulos, Karim Lasri, Naomi Saphra, Arabella Sinclair, et al.

State-of-the-art generalisation research in nlp: a taxonomy and review. *arXiv preprint* arXiv:2210.03050, 2022.

53. Frank Keller. Cognitively plausible models of human language processing. In *Proceedings of the 48th Annual Meeting of the Association for Computational Linguistics: Short Papers*, pages 60–67, 2010.

54. Roberto Navigli. Natural language understanding: Instructions for (present and future) use. In *IJCAI*, volume 18, pages 5697–5702, 2018.

55. Emily M. Bender and Alexander Koller. Climbing towards NLU: On meaning, form, and understanding in the age of data. In *Proceedings of the 58th Annual Meeting of the Association for Computational Linguistics*, pages 5185–5198, Online, July 2020. Association for Computational Linguistics. https://doi.org/10.18653/v1/2020.acl-main.463. https://aclanthology.org/2020.acl-main.463.

56. Mario Giulianelli, Jack Harding, Florian Mohnert, Dieuwke Hupkes, and Willem Zuidema. Under the hood: Using diagnostic classifiers to investigate and improve how language models track agreement information. In *Proceedings of the 2018 EMNLP Workshop BlackboxNLP: Analyzing and Interpreting Neural Networks for NLP*, pages 240–248, Brussels, Belgium, November 2018. Association for Computational Linguistics. https://doi.org/10.18653/v1/W18-5426. https://aclanthology.org/W18-5426.

57. Harm Brouwer, Francesca Delogu, Noortje J Venhuizen, and Matthew W Crocker. Neurobehavioral correlates of surprisal in language comprehension: A neurocomputational model. *Frontiers in Psychology*, 12: 615538, 2021.

58. Tal Linzen, Emmanuel Dupoux, and Yoav Goldberg. Assessing the ability of LSTMs to learn syntax-sensitive dependencies. *Transactions of the Association for Computational Linguistics*, 4:521–535, 2016. https://doi.org/10.1162/tacl_a_00115. https://aclanthology.org/Q16-1037.

59. Jaap Jumelet and Dieuwke Hupkes. Do language models understand anything? on the ability of LSTMs to understand negative polarity items. In *Proceedings of the 2018 EMNLP Workshop BlackboxNLP: Analyzing and Interpreting Neural Networks for NLP*, pages 222–231, Brussels, Belgium, November 2018. Association for Computational Linguistics. https://doi.org/10.18653/v1/W18-5424. https://aclanthology.org/W18-5424.

60. Rebecca Marvin and Tal Linzen. Targeted syntactic evaluation of language models. In *Proceedings of the 2018 Conference on Empirical Methods in Natural Language Processing*, pages 1192–1202, Brussels, Belgium, October-November 2018. Association for Computational Linguistics. https://doi.org/10.18653/v1/D18-1151. https://aclanthology.org/D18-1151.

61. Jon Gauthier, Jennifer Hu, Ethan Wilcox, Peng Qian, and Roger Levy. SyntaxGym: An online platform for targeted evaluation of language models. In *Proceedings of the 58th Annual Meeting of the Association for Computational Linguistics: System Demonstrations*, pages 70–76, Online, July 2020. Association for Computational Linguistics. https://doi.org/10.18653/v1/2020.acl-demos.10. https://aclanthology.org/2020.acl-demos.10.

62. Kristina Gulordava, Piotr Bojanowski, Edouard Grave, Tal Linzen, and Marco Baroni. Colorless green recurrent networks dream hierarchically. In *Proceedings of the 2018 Conference of the North American Chapter of the Association for Computational Linguistics: Human Language Technologies, Volume 1 (Long Papers)*, pages 1195–1205, New Orleans, Louisiana, June 2018. Association for Computational Linguistics. https://doi.org/10.18653/v1/N18-1108. https://aclanthology.org/N18-1108.

63. Jannis Vamvas and Rico Sennrich. On the limits of minimal pairs in contrastive evaluation. In *Proceedings of the Fourth BlackboxNLP Workshop on Analyzing and Interpreting Neural Networks for NLP*, pages 58–68, Punta Cana, Dominican Republic, November 2021. Association for Computational Linguistics. https://doi.org/10.18653/v1/2021.blackboxnlp-1.5. https://aclanthology.org/2021.blackboxnlp-1.5.

64. Hector Levesque, Ernest Davis, and Leora Morgenstern. The winograd schema challenge. In *Thirteenth international conference on the principles of knowledge representation and reasoning*, 2012.

65. Ernest Davis, Leora Morgenstern, and Charles L Ortiz. The first winograd schema challenge at ijcai-16. *AI Magazine*, 38 (3): 97–98, 2017.

66. Rachel Rudinger, Jason Naradowsky, Brian Leonard, and Benjamin Van Durme. Gender bias in coreference resolution. In *Proceedings of the 2018 Conference of the North American Chapter of the Association for Computational Linguistics: Human Language Technologies, Volume 2 (Short Papers)*, pages 8–14, New Orleans, Louisiana, June 2018. Association for Computational Linguistics. https://doi.org/10.18653/v1/N18-2002. https://aclanthology.org/N18-2002.

67. Vid Kocijan, Ernest Davis, Thomas Lukasiewicz, Gary Marcus, and Leora Morgenstern. The defeat of the winograd schema challenge. *arXiv preprint* arXiv:2201.02387, 2022.

68. Tal Linzen. How can we accelerate progress towards human-like linguistic generalization? In *Proceedings of the 58th Annual Meeting of the Association for Computational Linguistics*, pages 5210–5217, Online, July 2020. Association for Computational Linguistics. https://doi.org/10.18653/v1/2020.acl-main.465. https://aclanthology.org/2020.acl-main.465.

69. Alex Wang, Amanpreet Singh, Julian Michael, Felix Hill, Omer Levy, and Samuel Bowman. GLUE: A multi-task benchmark and analysis platform for natural language understanding. In *Proceedings of the 2018 EMNLP Workshop BlackboxNLP: Analyzing and Interpreting Neural Networks for NLP*, pages 353–355, Brussels, Belgium, November 2018. Association for Computational Linguistics. https://doi.org/10.18653/v1/W18-5446. https://aclanthology.org/W18-5446.

70. Alex Wang, Yada Pruksachatkun, Nikita Nangia, Amanpreet Singh, Julian Michael, Felix Hill, Omer Levy, and Samuel Bowman. Superglue: A stickier benchmark for general-purpose language understanding systems. *Advances in neural information processing systems*, 32, 2019.

71. Sebastian Ruder. Challenges and Opportunities in NLP Benchmarking. http://ruder.io/nlp-benchmarking, 2021.

72. Aarohi Srivastava, Abhinav Rastogi, Abhishek Rao, Abu Awal Md Shoeb, Abubakar Abid, Adam Fisch, Adam R Brown, Adam Santoro, Aditya Gupta, Adrià Garriga-Alonso, et al. Beyond the imitation game: Quantifying and extrapolating the capabilities of language models. *arXiv preprint* arXiv:2206.04615, 2022.

73. Lloyd S Shapley. A value for n-person games. *Classics in game theory*, 69, 1997.

74. Ian Covert, Scott M Lundberg, and Su-In Lee. Explaining by removing: A unified framework for model explanation. *J. Mach. Learn. Res.*, 22: 209–1, 2021.

75. Amit Dhurandhar, Pin-Yu Chen, Ronny Luss, Chun-Chen Tu, Paishun Ting, Karthikeyan Shanmugam, and Payel Das. Explanations based on the missing: towards contrastive explanations with pertinent negatives. In *Proceedings of the 32nd International Conference on Neural Information Processing Systems*, pages 590–601, 2018.

76. Sandipan Sikdar, Parantapa Bhattacharya, and Kieran Heese. Integrated directional gradients: Feature interaction attribution for neural NLP models. In *Proceedings of the 59th Annual Meeting of the Association for Computational Linguistics and the 11th International Joint Conference on Natural Language Processing (Volume 1: Long Papers)*, pages 865–878, Online, August 2021. Association for Computational Linguistics. https://doi.org/10.18653/v1/2021.acl-long.71. https://aclanthology.org/2021.acl-long.71.

77. Jaap Jumelet, Willem Zuidema, and Dieuwke Hupkes. Analysing neural language models: Contextual decomposition reveals default reasoning in number and gender assignment. In *Proceedings of the 23rd Conference on Computational Natural Language Learning (CoNLL)*, pages 1–11, Hong Kong, China, November 2019. Association for Computational Linguistics. https://doi.org/10.18653/v1/K19-1001. https://aclanthology.org/K19-1001.

78. Marco Tulio Ribeiro, Tongshuang Wu, Carlos Guestrin, and Sameer Singh. Beyond accuracy: Behavioral testing of NLP models with CheckList. In *Proceedings of the 58th Annual Meeting of the Association for Computational Linguistics*, pages 4902–4912, Online, July 2020. Association for Computational Linguistics. https://doi.org/10.18653/v1/2020.acl-main.442. https://aclanthology.org/2020.acl-main.442.

79. Koustuv Sinha, Prasanna Parthasarathi, Joelle Pineau, and Adina Williams. UnNatural Language Inference. In *Proceedings of the 59th Annual Meeting of the Association for Computational Linguistics and the 11th International Joint Conference on Natural Language Processing (Volume 1: Long Papers)*, pages 7329–7346, Online, August 2021. Association for Computational Linguistics. https://doi.org/10.18653/v1/2021.acl-long.569. https://aclanthology.org/2021.acl-long.569.

80. Eric Wallace, Pedro Rodriguez, Shi Feng, Ikuya Yamada, and Jordan Boyd-Graber. Trick me if you can: Human-in-the-loop generation of adversarial examples for question answering. *Transactions of the Association for Computational Linguistics*, 7:387–401, 2019. https://doi.org/10.1162/tacl_a_00279. https://aclanthology.org/Q19-1029.

81. Yuning Ding, Brian Riordan, Andrea Horbach, Aoife Cahill, and Torsten Zesch. Don't take "nswvtnvakgxpm" for an answer –the surprising vulnerability of automatic content scoring systems to adversarial input. In *Proceedings of the 28th International Conference on Computational Linguistics*, pages 882–892, Barcelona, Spain (Online), December 2020. International Committee on Computational Linguistics. https://doi.org/10.18653/v1/2020.coling-main.76. https://aclanthology.org/2020.coling-main.76.

82. Wencong You and Daniel Lowd. Towards stronger adversarial baselines through human-AI collaboration. In *Proceedings of NLP Power! The First Workshop on Efficient Benchmarking in NLP*, pages 11–21, Dublin, Ireland, May 2022. Association for Computational Linguistics. https://doi.org/10.18653/v1/2022.nlppower-1.2. https://aclanthology.org/2022.nlppower-1.2.

83. Robin Jia, Aditi Raghunathan, Kerem Göksel, and Percy Liang. Certified robustness to adversarial word substitutions. In *Proceedings of the 2019 Conference on Empirical Methods in Natural Language Processing and the 9th International Joint Conference on Natural Language Processing (EMNLP-IJCNLP)*, pages 4129–4142, Hong Kong, China, November 2019. Association for Computational Linguistics. https://doi.org/10.18653/v1/D19-1423. https://aclanthology.org/D19-1423.

84. Siwon Kim, Jihun Yi, Eunji Kim, and Sungroh Yoon. Interpretation of NLP models through input marginalization. In *Proceedings of the 2020 Conference on Empirical Methods in Natural Language Processing (EMNLP)*, pages 3154–3167, Online, November 2020. Association for Computational Linguistics. https://doi.org/10.18653/v1/2020.emnlp-main.255. https://aclanthology.org/2020.emnlp-main.255.

85. Pang Wei Koh and Percy Liang. Understanding black-box predictions via influence functions. In *International conference on machine learning*, pages 1885–1894. PMLR, 2017.

86. Xiaofei Sun, Diyi Yang, Xiaoya Li, Tianwei Zhang, Yuxian Meng, Qiu Han, Guoyin Wang, Eduard Hovy, and Jiwei Li. Interpreting deep learning models in natural language processing: A review. *arXiv preprint* arXiv:2110.10470, 2021.

87. Dheeru Dua, Yizhong Wang, Pradeep Dasigi, Gabriel Stanovsky, Sameer Singh, and Matt Gardner. DROP: A reading comprehension benchmark requiring discrete reasoning over paragraphs. In *Proceedings of the 2019 Conference of the North American Chapter of the Association for Computational Linguistics: Human Language Technologies, Volume 1 (Long and Short Papers)*, pages 2368–2378, Minneapolis, Minnesota, June 2019. Association for Computational Linguistics. https://doi.org/10.18653/v1/N19-1246. https://aclanthology.org/N19-1246.

88. Sebastian Ruder and Avi Sil. Multi-domain multilingual question answering. In *Proceedings of the 2021 Conference on Empirical Methods in Natural Language Processing: Tuto-

rial Abstracts, pages 17–21, Punta Cana, Dominican Republic & Online, November 2021. Association for Computational Linguistics. https://doi.org/10.18653/v1/2021.emnlp-tutorials. 4. https://aclanthology.org/2021.emnlp-tutorials.4.

89. Yiding Hao, Simon Mendelsohn, Rachel Sterneck, Randi Martinez, and Robert Frank. Probabilistic predictions of people perusing: Evaluating metrics of language model performance for psycholinguistic modeling. In *Proceedings of the Workshop on Cognitive Modeling and Computational Linguistics*, pages 75–86, Online, November 2020. Association for Computational Linguistics. https://doi.org/10.18653/v1/2020.cmcl-1.10. https://aclanthology.org/2020.cmcl-1.10.

90. Edwin Simpson and Iryna Gurevych. Scalable bayesian preference learning for crowds. *Mach. Learn.*, 109 (4): 689–718, apr 2020. ISSN 0885-6125. https://doi.org/10.1007/s10994-019-05867-2. https://doi.org/10.1007/s10994-019-05867-2.

91. Edwin Simpson and Iryna Gurevych. A Bayesian approach for sequence tagging with crowds. In *Proceedings of the 2019 Conference on Empirical Methods in Natural Language Processing and the 9th International Joint Conference on Natural Language Processing (EMNLP-IJCNLP)*, pages 1093–1104, Hong Kong, China, November 2019. Association for Computational Linguistics. https://doi.org/10.18653/v1/D19-1101. https://aclanthology.org/D19-1101.

92. Yarin Gal and Zoubin Ghahramani. Dropout as a bayesian approximation: Representing model uncertainty in deep learning. In Maria Florina Balcan and Kilian Q. Weinberger, editors, *Proceedings of The 33rd International Conference on Machine Learning*, volume 48 of *Proceedings of Machine Learning Research*, pages 1050–1059, New York, New York, USA, 20–22 Jun 2016. PMLR. https://proceedings.mlr.press/v48/gal16.html.

93. Francesco Verdoja and Ville Kyrki. Notes on the behavior of mc dropout. In *ICML Workshop on Uncertainty & Robustness in Deep Learning*, 2021.

94. Artem Shelmanov, Evgenii Tsymbalov, Dmitri Puzyrev, Kirill Fedyanin, Alexander Panchenko, and Maxim Panov. How certain is your Transformer? In *Proceedings of the 16th Conference of the European Chapter of the Association for Computational Linguistics: Main Volume*, pages 1833–1840, Online, April 2021. Association for Computational Linguistics. https://doi.org/10. 18653/v1/2021.eacl-main.157. https://aclanthology.org/2021.eacl-main.157.

95. Xiang Zhou, Yixin Nie, and Mohit Bansal. Distributed NLI: Learning to predict human opinion distributions for language reasoning. In *Findings of the Association for Computational Linguistics: ACL 2022*, pages 972–987, Dublin, Ireland, May 2022c. Association for Computational Linguistics. https://doi.org/10.18653/v1/2022.findings-acl.79. https://aclanthology.org/2022.findings-acl.79.

96. Alexios Gidiotis and Grigorios Tsoumakas. Should we trust this summary? Bayesian abstractive summarization to the rescue. In *Findings of the Association for Computational Linguistics: ACL 2022*, pages 4119–4131, Dublin, Ireland, May 2022. Association for Computational Linguistics. https://doi.org/10.18653/v1/2022.findings-acl.325. https://aclanthology.org/2022.findings-acl.325.

97. Burrhus F Skinner. Reinforcement today. *American Psychologist*, 13 (3): 94, 1958.

98. Kai A Krueger and Peter Dayan. Flexible shaping: How learning in small steps helps. *Cognition*, 110 (3): 380–394, 2009.

99. Natalie Schluter and Daniel Varab. When data permutations are pathological: the case of neural natural language inference. In *Proceedings of the 2018 Conference on Empirical Methods in Natural Language Processing*, pages 4935–4939, Brussels, Belgium, October-November 2018. Association for Computational Linguistics. https://doi.org/10.18653/v1/D18-1534. https://aclanthology.org/D18-1534.

100. Guy Hacohen and Daphna Weinshall. On the power of curriculum learning in training deep networks. In Kamalika Chaudhuri and Ruslan Salakhutdinov, editors, *Proceedings of the 36th International Conference on Machine Learning*, volume 97 of *Proceedings of Machine Learning Research*, pages 2535–2544. PMLR, 09–15 Jun 2019. https://proceedings.mlr.press/v97/hacohen19a.html.

101. Petru Soviany, Radu Tudor Ionescu, Paolo Rota, and Nicu Sebe. Curriculum learning: A survey. *International Journal of Computer Vision*, pages 1–40, 2022.

102. Benjamin Samuel Bloom. Taxonomy of educational objectives: The classification of educational goals. *Cognitive domain*, 1956.

103. Daniel Campos. Curriculum learning for language modeling. *arXiv preprint* arXiv:2108.02170, 2021.

104. Arielle Borovsky, Jeffrey L Elman, and Anne Fernald. Knowing a lot for one's age: Vocabulary skill and not age is associated with anticipatory incremental sentence interpretation in children and adults. *Journal of experimental child psychology*, 112 (4): 417–436, 2012.

105. Lucas Willems, Salem Lahlou, and Yoshua Bengio. Mastering rate based curriculum learning. *arXiv preprint* arXiv:2008.06456, 2020.

106. John P. Lalor and Hong Yu. Dynamic data selection for curriculum learning via ability estimation. In *Findings of the Association for Computational Linguistics: EMNLP 2020*, pages 545–555, Online, November 2020. Association for Computational Linguistics. https://doi.org/10.18653/v1/2020.findings-emnlp.48. https://aclanthology.org/2020.findings-emnlp.48.

107. Tom Kocmi and Ondřej Bojar. Curriculum learning and minibatch bucketing in neural machine translation. In *Proceedings of the International Conference Recent Advances in Natural Language Processing, RANLP 2017*, pages 379–386, 2017.

108. Cao Liu, Shizhu He, Kang Liu, Jun Zhao, et al. Curriculum learning for natural answer generation. In *IJCAI*, pages 4223–4229, 2018b.

109. Mrinmaya Sachan and Eric Xing. Easy questions first? a case study on curriculum learning for question answering. In *Proceedings of the 54th Annual Meeting of the Association for Computational Linguistics (Volume 1: Long Papers)*, pages 453–463, Berlin, Germany, August 2016. Association for Computational Linguistics. https://doi.org/10.18653/v1/P16-1043. https://aclanthology.org/P16-1043.

110. Alexis Conneau and Guillaume Lample. Cross-lingual language model pretraining. *Advances in neural information processing systems*, 32, 2019.

111. Pratik Joshi, Sebastin Santy, Amar Budhiraja, Kalika Bali, and Monojit Choudhury. The state and fate of linguistic diversity and inclusion in the NLP world. In *Proceedings of the 58th Annual Meeting of the Association for Computational Linguistics*, pages 6282–6293, Online, July 2020. Association for Computational Linguistics. https://doi.org/10.18653/v1/2020.acl-main.560. https://aclanthology.org/2020.acl-main.560.

112. Laura Cabello Piqueras and Anders Søgaard. Are pretrained multilingual models equally fair across languages? In *Proceedings of the 29th International Conference on Computational Linguistics*, pages 3597–3605, Gyeongju, Republic of Korea, October 2022. International Committee on Computational Linguistics. https://aclanthology.org/2022.coling-1.318.

113. Davis Liang, Hila Gonen, Yuning Mao, Rui Hou, Naman Goyal, Marjan Ghazvininejad, Luke Zettlemoyer, and Madian Khabsa. Xlm-v: Overcoming the vocabulary bottleneck in multilingual masked language models, 2023.

114. Saima Malik-Moraleda, Dima Ayyash, Jeanne Gall?e, Josef Affourtit, Malte Hoffmann, Zachary Mineroff, Olessia Jouravlev, and Evelina Fedorenko. An investigation across 45 languages and 12 language families reveals a universal language network. *Nature Neuroscience*, 25: 1–6, 08 2022b. https://doi.org/10.1038/s41593-022-01114-5.

115. Elisabeth Norcliffe, Alice C. Harris, and T. Florian Jaeger. Cross-linguistic psycholinguistics and its critical role in theory development: early beginnings and recent advances. *Language, Cognition and Neuroscience*, 30 (9): 1009–1032, 2015. https://doi.org/10.1080/23273798. 2015.1080373. https://doi.org/10.1080/23273798.2015.1080373.

116. Daniel Kahneman. A perspective on judgment and choice: mapping bounded rationality. *American psychologist*, 58(9):697, 2003.

117. Emiel van Miltenburg. Stereotyping and bias in the flickr30k dataset. In *Proceedings of Multimodal Corpora: Computer vision and language processing (MMC 2016)*, pages 1–4. 2016.

118. Emily M. Bender, Dirk Hovy, and Alexandra Schofield. Integrating ethics into the NLP curriculum. In *Proceedings of the 58th Annual Meeting of the Association for Computational Linguistics: Tutorial Abstracts*, pages 6–9, Online, July 2020. Association for Computational Linguistics. https://doi.org/10.18653/v1/2020.acl-tutorials.2. https://aclanthology.org/ 2020.acl-tutorials.2.

119. Ninareh Mehrabi, Fred Morstatter, Nripsuta Saxena, Kristina Lerman, and Aram Galstyan. A survey on bias and fairness in machine learning. *ACM Comput. Surv.*, 54 (6), jul 2021. ISSN 0360-0300. https://doi.org/10.1145/3457607. https://doi.org/10.1145/3457607.

120. Tom McCoy, Ellie Pavlick, and Tal Linzen. Right for the wrong reasons: Diagnosing syntactic heuristics in natural language inference. In *Proceedings of the 57th Annual Meeting of the Association for Computational Linguistics*, pages 3428–3448, Florence, Italy, July 2019. Association for Computational Linguistics. https://doi.org/10.18653/v1/P19-1334. https://aclanthology. org/P19-1334.

121. Olivia Guest and Andrea E Martin. On logical inference over brains, behaviour, and artificial neural networks. *Computational Brain & Behavior*, pages 1–15, 2023.

122. Anders Søgaard. Explainable natural language processing. *Synthesis Lectures on Human Language Technologies*, 14(3):1–123, 2021.

Representational Structure

5

When we are examining the input–output patterns of a model, we can discover recurring patterns in its behavior and infer strategies that the model might have learned, but we can only speculate why certain patterns occur. For example, if a model performs well on number agreement, it remains an open question whether it has acquired a rule that an English singular noun in the subject needs to be matched with a main verb in the corresponding singular form or if it has simply learned lexical co-occurrences (e.g., that *cat* co-occurs more often with *purrs* than with *purr*). We cannot even be sure if the model establishes the link between singular and plural occurrences of the same noun (*cat* and *cats* might be represented just as unrelated as *cat* and *newspaper*). In order to better understand how a model encodes the input and organizes knowledge, we need to have a closer look at its internal representations.

In this chapter, we describe how we can analyze the representational structure of a model by examining similarity relations in the representational space. We discuss the criteria for comparing the distributional spaces and how these representations can be tested through probing methods. Finally, we argue that grounding language in multimodal aspects by integrating cognitive signals into the models.

5.1 Analyzing Representational Structure

In neural language models, the input is represented in high-dimensional vector space and the representations are optimized with respect to a pretraining objective (see Chap. 2). Current training objectives are based on the distributional hypothesis which postulates that relations between tokens can be determined according to the contexts in which they appear [1, 2]. According to this usage-based understanding of language, it is not possible to identify absolute truths about what a concept means or how a linguistic construction. Symbolic mappings and combinatory rules are arbitrary agreements that can be changed if a group of humans

© The Author(s), under exclusive license to Springer Nature Switzerland AG 2024   89
L. Beinborn and N. Hollenstein, *Cognitive Plausibility in Natural Language Processing*, Synthesis Lectures on Human Language Technologies,
https://doi.org/10.1007/978-3-031-43260-6_5

decides to use them differently. According to this more fluid notion of meaning, the quality of vector representations for language input is not determined by the absolute positions of the vectors but by their relative distribution in the vector space. Within a representational space, relations between instances are inferred from their representational similarity [3].

5.1.1 Representational Similarity

Distributional representations of language are usually developed for the following goal: similar inputs should be mapped to vectors that are close to each other in vector space. Mathematically, the most common metric to quantify the representational distance between two vectors **x** and **y** is the cosine similarity. It measures the cosine angle between the two vectors by taking the dot product between the vectors normalized by their length:

$$cos(\mathbf{x}, \mathbf{y}) = \frac{\mathbf{x} \cdot \mathbf{y}}{|\mathbf{x}| \cdot |\mathbf{y}|} \tag{5.1}$$

The resulting value is close to 1 if the vectors point in the same direction, and close to -1 if the vectors point in the opposite direction. It can be used directly for expressing similarity or by subtracting it from 1 for expressing distance.

5.1.1.1 Similarity between Words

Conceptually, the similarity between two language inputs is an underspecified phenomenon. Even if we restrict the input to refer to a pair of isolated words, it is difficult to quantify the similarity between them reliably. Whether we perceive two words to be similar strongly depends on the lens through which we examine them and on the frame of reference that we use as a contrast. The maximum for similarity is usually determined by synonyms that are semantically interchangeable in most contexts (*teacher* and *instructor*). For English, the gold standard for highly similar concepts is usually determined by the *synset* relations encoded in the lexical resource WordNet [4]. More graded distinctions of similarity require very elaborate definitions and an illustration of edge cases to determine a reliable scale [5]. For example, two words might be conceptually related (a *bucket* and a *wallet* both usually contain something), taxonomically related (a *cucumber* and an *onion* are both labeled as vegetables in biological taxonomies), syntactically related (*flew* and *drank* are both irregular past tense verbs), associatively related (*confetti* and *champagne* are both used for parties), or emotionally related (*clowns* and *spiders* can both be perceived as scary). The two words might rhyme or be associated with similar colors, tastes, or shapes. Sometimes two concepts are perceived to be related to each other because they are both linked to a current event or news topic (e.g., *restaurant* and *vaccination* became related during the pandemic). Even antonyms can be perceived as similar, for example, both *curly* and *straight* refer to types of hair. In order to narrow down these concepts, a distinction between semantic similarity and relatedness (or association) has been proposed [6, 7]. Hill et al. [8] define semantic similarity as an overlap in properties based on the feature-based model by Tversky [9].[1] They find that distributional models struggle with this restricted concept of similarity and are better at representing more associative relations between words.

Language models integrate a multi-faceted perspective on the relative similarity between inputs because representational similarity is assessed over a large number of dimensions. When language models are finetuned to a specific task, the representations of tokens move within the vector space. This can be conceptually interpreted as prioritizing certain dimensions over others.

5.1.1.2 Contextualization

In contextualized models, the representation of an individual token changes with respect to the context in which it occurs to account for polysemy. Accordingly, its relation to other tokens is not static but varies with respect to a larger input sequence. Most language models stack multiple representational layers on top of each other to enable hierarchical processing. The representation of the input varies across layers and studies that assess the self-similarity of token representations across context indicate that contextualization increases in higher layers [10, 11]. Assessing the self-similarity of token representations requires an alignment of representations to input tokens. Brunner et al. [12] question whether this is conceptually meaningful because self-attention mixes the input information by aggregating context into the hidden representations in higher layers. They find that the local context is progressively aggregated into the vector but that the original input token is still identifiable. Ethayarajh [10] finds that the representation of functional words like articles and prepositions is surprisingly context-specific. He concludes that the contextualization of a token is driven by the variety of contexts it appears in, rather than by the number of different senses it can have. This finding supports the assumption that representational similarity in contextualized models rather represents associative relations than semantic similarity.

If we move beyond words, the similarity between inputs becomes more complex. Contextualized language models can encode full sentences but which criteria do we apply to assess their similarity? Commonly, two sentences are considered to be similar if they are paraphrases [13]. In a series of SemEval tasks, Agirre et al. [14] aim at more graded distinctions and evaluate the similarity between two sentences on a 6-point Likert scale. For example, they distinguish whether two sentences "are mostly equivalent, but some unimportant details differ", "are roughly equivalent, but some important information differs/missing", "are not equivalent, but share some details", "are not equivalent, but are on the same topic". They provide examples for each category but it remains unclear how to interpret these categories in practice. Are a sentence in present tense and its equivalent in past tense more similar than a sentence and an approximate paraphrase of it in jargon? Is a sentence similar to its negated version or are the two strongly dissimilar?

---

[1] This shifts the challenge of defining similarity to defining what qualifies as a property.

Some contextualized language models jointly optimize a dedicated sentence representation using the next-sentence prediction objective (see Chap. 2). This *CLS* token is often finetuned for sentence classification but Reimers and Gurevych [15] find that taking the mean over the token representations is a better approximation for sentence similarity. Ethayarajh [10] calculates the similarity between tokens of a sentence to its mean representation and compares the behavior over different layers. They analyze English ELMo, BERT, and GPT-2 models and find that the tendencies for intra-sentence similarity differ strongly. The three models optimize different target objectives but they also differ in the model architecture and the training data which makes it difficult to infer the source of these differences.

If we zoom in on subwords which are the input tokens currently used by most neural language models, it becomes impossible to conceptually interpret the similarity between tokens because many subwords cannot be mapped onto linguistically meaningful units.[2] Many token-level analyses of representational similarity therefore either ignore input tokens that are split into subwords or sum or average the vector representations of the parts into a single representation [16, 17].

5.1.1.3 Representational Clusters

Mimno and Thompson [18] and Ethayarajh [10] examine representational structure from a general geometric perspective. They find that the representations in most language models are organized in an anisotropic structure that uses only a relatively small cone-shaped manifold in the representational space. As a consequence, even pairs of randomly chosen words are representationally similar and are assigned a high cosine similarity. Cai et al. [19] extend this study and find that the anisotropic structure corresponds to a global view. When zooming in on clusters in the vector space, more isotropic distributions can be observed.

Papernot and McDaniel [20] take a task-specific perspective to analyze how instances are grouped in the representational space. They apply a k-nearest neighbors interpretation to determine which group of training instances is relevant for the prediction of a test instance. For this approach, the representation of the test instance is compared to the representation of the training instances in every layer of the model. The conformity score then indicates the ratio of k-nearest neighbors belonging to the same class. It estimates to which extent the representation of the training data supports a classification decision and can be interpreted as the model's confidence [21]. Groups of instances with high conformity indicate that the representational cluster captures information that correlates with task-specific knowledge.

The knowledge that is shared within a group of instances can be characterized by the concept activation vector [22]. For this approach, a linear classifier is trained to distinguish between a group of instances sharing a concept and a group of random counter-examples. The vector orthogonal to the decision boundary of the classifier represents the concept vector, a prototypical representation of the concept. For a test instance, we can then analyze how the model's decision would change if the instance were more prototypical, i.e., if it would

---

[2] Recall the example of *sunscreen* in Chap. 2 which is split into *[sun, ##sc, ##reen]*.

be manipulated towards the direction of the concept vector, by calculating the directional derivatives. Such continuous modifications can provide intuitive interpretations for images because they correspond to changes in pixel values. For example, an image representation of a dog can be modified to be closer to the concept vector for *striped*. While one might have ideas about how to modify a sentence to be closer to the stylistic concept *academic* or to the linguistic concept *passive* by changing words, these symbolic changes cannot be directly mapped into continuous vector modifications [23].

5.1.2 Comparing Representational Spaces

The cognitive plausibility of distributional representations is traditionally assessed by the quality of the encoded relations. For a subset of pairs of inputs, the distances in vector space are compared to relational judgments obtained from humans. Reimers et al. [24] show that the commonly used Pearson's correlation coefficient is not necessarily the most predictive metric for measuring similarity. They propose a set of alternative metrics including ranking and accuracy measures that should be chosen with respect to the downstream task requirements. They strongly recommend that the instance-level similarity scores should always be made available for better transparency of the results (see also Sect. 4.1.2.2).

5.1.2.1 Psycholinguistic Criteria

For English representations, the similarity between tokens is commonly tested using similarity datasets such as SimLex999 [25], MEN [26], WS353 [27], RW [28], and SemEval-2012 [29]. For these datasets, annotators are asked to rate the similarity or relatedness of pairs of words on a Likert scale. The granularity of the annotation guidelines differs and the conceptual setup of similarity studies has been criticized because it leads to low inter-annotator agreement and ignores well-known psychological phenomena such as the asymmetry of similarity judgments [9, 30]. On the sentence level, several similarity datasets have been assembled in the test suite *SentEval* [31] to compare the representational quality of sentences from multiple perspectives. Many similarity datasets have been translated into other languages but it is questionable whether similarity values could be propagated across languages. As polysemy varies across languages, similarity judgments of polysemous inputs will likely vary depending on the translation [32].

In an attempt to narrow down the set of properties on which similarity should be evaluated, odd-one-out tasks evaluate groups of concepts. For this task, humans are asked to identify a word that stands out from a group of four or five words because it does not have a property that is shared by the other words. For example, out of the four words *sandwich, soup, plate*, and *cake*, *plate* would be the odd-one-out because it is not edible. Unfortunately, for many odd-one-out examples, the relevant property can often not be unambiguously detected because different groups can be identified depending on the perspective [33]. For example, one could also choose *soup* as the odd one out for being liquid.

Pairwise analogies characterize the perspective through which similarity is assessed more explicitly [34]. Turney [35] defined a pair of words to be analogous to another pair when the semantic relations between the words in the first pair are highly similar to the relations in the second pair. In analogy tasks, an example is provided and the goal is to find a similar relation for a stimulus. For example, for the prompt *The relation between pretty and ugly is similar to the relation between big and X*, we would expect antonyms of *big* such as *small* or *tiny* as candidates for *X*. If we instead ask for the relation between *pretty* and *prettier*, a morphological perspective is applied and we would expect the comparative *bigger* as a solution for *X*. In the original setup, a single solution is expected for *X* and the other three concepts in the analogy are excluded from the set of candidates. In practice, human responses to analogy solving are subject to biases and can be influenced by priming [36].

To go beyond the concept of token-level similarity, the representational structure is often compared to psycholinguistic categories such as frequency, concreteness, age of acquisition, and imageability. For English, these values can be extracted from the MRC database [37]. Rogers et al. [38] claim that a binary understanding of linguistic relations is an oversimplification and that a cognitively more plausible evaluation of representational structure should account for the graded nature of relational similarity in human reasoning.

5.1.2.2 Cognitive Similarity

The conscious annotation decisions for similarity are affected by the criteria described in the annotation guidelines [25]. Cognitive signals recorded during language processing make it possible to tap more directly into the subconscious representational structure underlying human language understanding.

As described in Chap. 3, response times for lexical decision tasks can be influenced by semantically related priming stimuli. Ettinger and Linzen [39] examine whether this relationship can be approximated by computational representations of the stimuli. They calculate the similarity between the representations and fit a linear model to predict the response times of the SPP dataset [40]. Auguste et al. [41] build on this finding and develop an evaluation framework that calculates the cognitive plausibility of model representations as the Spearman correlation between the response times and the cosine similarity between the computational representations of the respective stimuli. They find that Glove representations correlate better with the response times than explicit human ratings for word similarity tasks.

FMRI scans provide a spatially distributed representation of neuronal activity in the brain (see Chap. 3). For visual processing, Kriegeskorte et al. [42] coined the term representational similarity analysis (RSA) to compare relations between high-dimensional vectors across representational spaces. In this case, the cosine similarity between two vectors in the representational space of the model is not directly compared to a one-dimensional similarity value but to the similarity between a pair of corresponding vectors obtained from another high-dimensional representational space. For example, the vectors for the concepts *bike* and *car* are expected to be more similar to each other than *bike* and *ant* both in the brain activation signal and in the language model.

5.1.2.3 Comparing Models

The strategy of second-order comparison is not only useful for comparing models and brain activation signals but also for comparing the representational structure across models [43–46]. We can systematically track changes in the relative structure with respect to the model architecture or the training objective. From a cognitive perspective, comparisons between different modalities have been very popular recently as purely text-based models lack perceptual grounding in the world.

When comparing representational structure across models, it is important to control for spurious correlations in the high-dimensional data. Both language model representations and cognitive signals exhibit anisotropic structure [10, 47]. This indicates that truly random vectors are not a good basis for comparison as they do not follow comparable distributional patterns. Instead, a permutation baseline that randomly reassigns stimuli to a non-matching representation (both cognitive and computational) provides a more robust control condition [48, 49].

To quantify the effect of enriching a language model with cognitive data (see Sect. 5.3.2), it is common practice to compare the performance to a purely text-based language model to show the improvement gained from the additional information available in the cognitive signals [50]. While this evaluation setup is intuitively sensible, the performance gain might simply be attributed to the increased expressive power of the model due to a larger number of input dimensions. It should therefore be combined with the permutation approach that ignores the alignment between stimuli and cognitive responses and adds the cognitive signals in a randomized fashion. When working with eye-tracking signals, a meaningful comparison involves the integration of word length and word frequency information [51, 52] as eye-tracking responses are known to be sensitive to these properties.

When calculating statistical significance for such analyses, one needs to correct for the multiple-hypotheses problem. This occurs when several individual hypothesis tests are considered simultaneously, e.g., when comparing various model configurations, multiple datasets, or the impact of various types of cognitive signals. In the case of a multiple-hypotheses problem, the significance or the error rate of individual tests no longer represents the error rate of the combined set of tests. Testing methods for multiple-hypotheses setups need to correct error rates accordingly [53]. A common strategy to rectify this problem is the application of the Bonferroni correction [54], which controls for the family-wise error rate (FWER). Under the Bonferroni correction, the global null hypothesis is rejected if $p < \alpha/N$, where N is the number of hypotheses. See Dror et al. [55] for more details about applying the Bonferroni correction in NLP models. Other statistical correction methods control the false discovery rate [56].

Most representational analyses heavily rely on a geometrical distance metric and commonly use cosine similarity. Recent studies indicate that cosine similarity should be interpreted with caution in contextualized representations [57]. More specifically, Timkey and van Schijndel [58] find that cosine similarity is overly sensitive to spurious dimensions and find "a striking mismatch between the dimensions that dominate similarity measures and those which are important to the behavior of the model." Zhou et al. [59] show that more frequent words are most affected by this behavior.

The quality of the representational structure is often assessed by the task-specific performance of a finetuned model. As the tested models usually vary in a multitude of experimental parameters (e.g., size and quality of training data, model architecture, training parameters) it is usually not possible to quantify the influence of the representational structure on the performance differences alone. Novel interpretability techniques can provide complementary insights into the role of the underlying representations.

5.2   Testing Representational Characteristics

When we analyze the relative structure of distributed representations, we build up ideas about how concepts are represented by the model in an exploratory fashion. If we want to go beyond such anecdotal findings, we need to develop specific hypotheses about the representational structure which can be explicitly tested. For evaluating the cognitive plausibility of the representations, we aim at developing hypotheses that link cognitively grounded findings about human language processing and assumptions about the model architecture.

Probing classifiers have become a popular method for evaluating hypotheses about the knowledge encoded in the hidden representations of neural language models. Probing classifiers take the representation of the model as an input and learn to classify a certain characteristic from this representation without fine-tuning the input representation [60].

Probing tasks have the structure $F:X \rightarrow Y$. Given a set of representations X, we want to find the best mapping F that relates X to a set of target features Y using a supervised model. In this section, we distinguish between probing for linguistic knowledge and probing for brain activation patterns.

5.2.1   Probing Linguistic Knowledge

Linguistic probing is based on the intuition that a neural model encodes a linguistic property if a classifier can learn to accurately predict the property from the representations of the model. Probing classifiers have been used to investigate whether language models encode morphological knowledge such as *person* and *number* features [61, 62], syntactic knowledge such as word classes [63, 64], anaphora resolution [65] or incremental syntactic states [66], and semantic knowledge about word senses [67]. The diagnostic datasets introduced in Sect.

4.2.1 are often used for probing representational structure because they are carefully designed to control for linguistic phenomena [68]. Conneau and Kiela [31] assembled SentEval, a toolkit to evaluate sentence representations including probing tasks for word order, sentence length, tree depth, and coordination inversion. Rogers et al. [69] provide an overview of probing tasks applied to the BERT model.

5.2.1.1 Task-Specific Knowledge

Rogers et al. [70] take a more applied perspective and probe for task-specific knowledge about named entity recognition, semantic role labeling, relation detection, and co-reference resolution. They find that contextualized models are more successful in encoding knowledge required for syntactic tasks than a non-contextual baseline. The differences for semantic tasks are considerably smaller indicating that contextualization contributes to encoding structural knowledge rather than deeper semantic compositionality (see also Sect. 5.1.1.2). Conneau et al. [71] point out that applied downstream tasks require complex combinations of cues and advocate for one-dimensional probing tasks to isolate properties and better control for biases. They provide word content recovery as an example of such a simpler task. They decode individual words from representations of sentences containing the word and show that sentence representations which conserve access to the lexical components perform better on most downstream tasks. This supports the assumption that reliance on lexicalized patterns is often a successful shortcut for good task performance and that more high-level abstractions are not necessary if the datasets are not balanced (see Sect. 4.2.2).

Many natural language processing tasks require commonsense knowledge that is often expressed as relations between concepts. Commonsense knowledge bases such as Concept-Net encode, for example, the typical location of a concept (the *savannah* for a lion, the *sett* for a badger, the *kitchen* for a microwave) but also more factual knowledge such as the capital of a country [72]. Recent studies show that commonsense relations can also be extracted from language models [73, 74]. As relational knowledge is usually expressed in triples of the form <head, relation, tail> (for example, <Rome, capital-of, Italy>), it cannot directly be probed from the language model. Instead, the relations are transformed into prompting templates ("The capitol of [MASK] is Rome") with either the head or the tail being masked and the predictions are ranked to evaluate the ability of the model to encode commonsense knowledge. The results by Petroni et al. [74] show that neural language model outperform more traditional approaches for extracting commonsense knowledge but the performance varies for different relation types. For instance, Lin et al. [75] find that language models perform poorly in reasoning about numerical commonsense knowledge (e.g., *A bird usually has [MASK] legs*). We think that it is important to distinguish between examples of commonsense knowledge that are essential to cognitively plausible reasoning and encyclopedic facts (e.g., a politician's year of birth) that might be relevant for task-specific applications but are not common knowledge for most humans. It is also important to be aware of the fact that models are not consistent with respect to factual knowledge and tend to contradict themselves [76].

5.2.1.2 Methodological Challenges

While the implementation of probing tasks is relatively straightforward, it is important to consider various methodological challenges before making experimental choices. It starts with the distinction between diagnostic probing and task-specific classification which is a blurry one. Generally, probing is understood as training a classifier that takes the unchanged model representations as input to predict a property. In contrast, task-specific finetuning directly modifies the representations of the model.

With respect to the choice of the probing classifier, it remains an open question how to determine a suitable trade-off between simplicity and expressivity [77]. Hupkes et al. [78] propose to restrict the probing classifier to a linear operation to focus on the knowledge already encoded in the representations and not on the general learnability of the problem. Hewitt and Liang [79] define the term *selectivity* to quantify the ability of a probe to distinguish linguistic abilities from simple memorization. In contrast, Pimentel et al. [80] point out that the distinction between identifying linguistic structure and learning the task is not warranted from an information-theoretic perspective. They discuss that it would be artificial to isolate the generalization and memorization competencies of a model from another because both play a key role in language understanding. They propose to always select the best-performing probe to capture potential non-linear relationships but at the same time, they cast doubt on the conclusions that can be drawn from such a setup.

Probing classifiers analyze *if* linguistic knowledge is encoded in a specific layer representation of the model. Torroba Hennigen et al. [81] call this approach extrinsic probing and introduce intrinsic probing which is targeted at further narrowing down *how* linguistic information is structured within a representation. They build on neuron-level probing to attribute conceptual knowledge or functional roles to individual neurons [82, 83]. As the strength of neural networks lies in the power of distributed knowledge that is redundantly organized [84], they propose to identify groups of neurons, for example, by selecting subsets of dimensions.

The probing classifier and the original model from which the representations are extracted are disconnected because they are trained in separate processes. Therefore, probing experiments can indicate correlations between the representations of a model and a specific linguistic property, but we cannot infer whether this property is involved in the predictive processes of the model. Elazar et al. [85] propose to measure the influence of a causal intervention that removes a property from the representation using iterative nullspace projection [86]. They call this process amnesic probing and show that conventional probing performance is not correlated to task importance in their experiments. These findings indicate that representational analyses of static frozen representations do not sufficiently capture the more fluid procedural dynamics in contextualized language models.

5.2.2   Probing Brain Activation Patterns

When probing for linguistic structure, we need to take a stance on what type of knowledge we consider crucial for language understanding and explicitly define expectations for cognitively plausible representations. However, the representational structure underlying language processing in the brain is also not yet understood. As an explorative shortcut, we can keep the concept of cognitive plausibility more implicit and directly probe computational representations for their ability to predict brain activation patterns. The underlying hypothesis is that similar words trigger similar activation patterns in the brain. If we can identify a mapping function from computational representations to brain activations, we are one step closer to pinning down the details of this similarity relation. While experiments with humans are heavily constrained due to ethical reasons, we can freely manipulate the experimental parameters of the computational model.

5.2.2.1   Encoding and Decoding

In cognitive neuroscience, mapping functions between brain responses and computational models are categorized as encoding or decoding operations [87]. Decoding aims at mapping brain activation patterns to the stimuli that generated them. This approach is often used to better understand how humans process language and how semantic representations are spatially distributed in the brain [88]. Murphy et al. [89] provide an overview of decoding research from a neuropsychological perspective. Gauthier and Ivanova [90] question the informativeness of encoding and decoding studies because differences between models are not transparently quantified. They propose to explicitly measure explained variance to better account for the noise in the brain response. More recently, Pascual et al. [91] propose a decoding approach that directly maps an fMRI scan to the corresponding word within a small fixed vocabulary. Unlike existing work, they evaluate their model with previously unseen participants. However, the top-1 accuracy of their approach is still <3% and highlights the need for further research. In this book, we focus on the reverse task known as encoding. The goal is to evaluate the cognitive plausibility of language models by measuring how relations between stimuli in the computational vector space of the model correspond to relations between stimuli in the brain activation patterns. We can rephrase this idea in the probing context as follows: if a classifier can learn to predict brain activation patterns from the representations of a neural model, then the model encodes information about language in a way that is similar to human language processing.

5.2.2.2   Word-level Probing

Mitchell et al. [92] initiated this line of research by recording fMRI brain activity data from participants who were instructed to think about the properties of a concrete noun such as *bear* or *refrigerator*. They map each stimulus into a vector representation based on semantic features extracted from a large corpus. They then train a linear classifier to predict the response captured in the fMRI scan from the feature representation. This approach assumes that the brain activity observed when thinking about a concrete noun can be derived as a weighted linear sum of contributions from each of its semantic features. As the fMRI dataset is publicly available, it has been used to evaluate the cognitive plausibility of different computational word representations including information from lexical resources, distributional, and neural approaches [93–96].

Hollenstein et al. [97] provide a cognitive evaluation framework to compare computational models by their ability to predict word-level eye-tracking or brain activity data. They find that the probing results are strongly correlated not only across types of cognitive processing signals but also between datasets of the same type. Their results show that models that better predict cognitive signals also yield better performance in two downstream language processing tasks. Fyshe et al. [98] and Athanasiou et al. [11] go one step further and directly integrate information about neural similarity into computational models to improve performance in downstream tasks.

Humans acquire word meaning in a multimodal environment and it is likely that brain activation patterns reflect multimodal properties whereas language models are commonly trained only on texts. Analyses with different models indicate that the integration of visual characteristics is particularly advantageous for computational representations of conrete words and representations derived from purely textual data are better at reflecting relations between abstract concepts [8, 26, 99, 100]. Linking these hypotheses to brain activation patterns leads to inconsistent results. Bulat et al. [101] find that visual representations yield better probing results for brain activation patterns of concrete concepts, while Anderson et al. [102] do not find a difference between modalities.

5.2.2.3 Towards Contextualized Analyses

The probing experiments described above are limited to isolated words. Hendrikx and Beinborn [103] point out that human concept representations are more fluid than dichotomous categories can capture as the context determines whether a word is experienced as more abstract or more concrete. For example, *tree* can refer to a concrete plant or to an abstract programming structure but both usages are conceptually related. This fluidity cannot be captured by probing analyses using static non-contextualized representations [104]. Pereira et al. [105] provide a dataset that accounts for the influence of context by recording fMRI signals in varying context paradigms: they represent the target concept within a sentence, alongside a picture, and surrounded by a word cloud of semantically related words.

The probing experiments have been extended to predicting brain responses recorded during reading longer sequences of naturally occurring stimuli [106] and across multiple languages [107]. In contextualized scenarios, additional modeling choices need to be made to account for the hemodynamic delay and spill-over effects. These choices have not yet been standardized and both the experimental setup and the evaluation protocol differ across contextualized probing studies which makes it difficult to generalize the results [108]. More recent studies aim at a more robust experimental setup and probe a wider range of model architectures for multiple types of cognitive signals. Toneva and Wehbe [109] use both fMRI and MEG signals and compare probing results for four language models to study the effects of layer depth, context length, and attention type on cognitive plausibility. Schrimpf et al. [110] run probing analyses on a range of contextualized transformer-based language models using datasets containing fMRI, ECoG, and self-paced reading signals. They find that the cognitive plausibility of a model seems to be closely linked to its performance on the next-word prediction task but do not isolate this experimental parameter explicitly. These contextualized probing studies are closely linked to the following chapter as they provide insights into procedural aspects of language. We think that the hemodynamic delay and the large individual differences in fMRI response patterns impede a more finegrained analysis of procedural strategies and focus on eye-tracking and EEG data in Chap. 6.

Linking the cognitive principles underlying reading and listening will be an important challenge toward better understanding language representations. A discussion of speech comprehension unfortunately remains out of the scope of this book but we would like to refer interested readers to contextualized probing studies using cognitive signals of listening [88, 111–116].

5.3 Towards Cognitively Plausible Representations

Most of the computational approaches described in this book focus on processing written texts. Neural language models interpret language as a continuous string of characters or tokens and follow a self-supervised training paradigm to reduce the need for external knowledge. From a cognitive perspective, these purely text-driven approaches only capture a small non-representative subset of language processing. Language is a means of synchronous communication that is grounded in perceptual experience and sensorimotor interactions with the physical world [117, 118].

A language model can only learn from knowledge that is encoded in its representations. If we abstract from the font type during training, we cannot use it as a cue for identifying irony expressed by the use of `comic sans`. In order to obtain cognitively more plausible representations of language, an important aspect that is currently being rediscovered in the field is the aim of grounding language. We first discuss methods for multimodal grounding in general and then zoom in on approaches that ground language by integrating cognitive signals as an additional modality.

5.3.1 Multimodal Grounding

In cognitive linguistics, the idea that language processing is not an isolated phenomenon but is influenced by our individual bodily experiences is known as *embodiment*. This has been strongly substantiated by empirical evidence and according to Bergen [119], it is "now inconceivable to most cognitive scientists that language, including syntax, could be informationally encapsulated, or that language wouldn't use other systems, including of the brain and body, or that individual experience wouldn't matter" (p.31). This individualized and embodied view on language processing has been largely ignored in the development of natural language processing models. Current models require large amounts of standardized data that rather represent a generic account of language knowledge that abstracts from individual experiences.

Research on representational structure often focuses on conceptual knowledge and the representation of static properties. Contextualized language models are a step forward towards a more situated model of language understanding because they make it explicit that meaning is determined by context. Research on multimodal semantics indicates that limiting the information to textual context alone favors representational structures that reflect taxonomical relations between concepts and their abstract properties because information about concrete perceptual properties regarding the shape, size, color, smell, or taste of concepts is underrepresented in texts [120–122]. We would rarely find references to *yellow bananas* in text because it is a redundant perceptual specification, but more abstract taxonomic distinctions such as *organic fruits* are quite frequent. Baroni [123] propose that combining multimodal information is a necessary step towards grounding representations of language in the world.

5.3.1.1 Multimodal Integration

Beinborn et al. [100] discuss different methods for combining information from multiple modalities. For multimodal fusion and multimodal mapping techniques, unimodal representations are first learned separately and are then combined into a shared representational space using dimensionality reduction or projection. With the increasing availability of image, audio, and video data and more efficient computational processing strategies, joint multimodal modeling is on the rise which aims at directly learning a representation that accommodates input from multiple modalities simultaneously.

Most multimodal approaches are application-oriented and provide models for tasks that are inherently multimodal such as image caption generation [124], visual question answering [125], or emotion detection [126]. See Baltrušaitis et al. [127] for an overview from a machine learning perspective.

In order to develop more generic, contextualized, and task-agnostic multimodal representations, transformer-based architectures that are successful in text-based language modeling have been combined with multimodal training objectives such as image-text matching or word-region alignment [128–131]. The overview by Pezzelle et al. [132] confirms that these models still struggle to integrate complementary multimodal information instead of simply reinforcing redundant cues, as observed earlier by Beinborn et al. [100] for static multimodal models. Novel multimodal probing tasks and diagnostic datasets are targeted at quantifying to which extent the modalities complement each other [125, 133–135]. Future research on cognitively plausible representations needs to go beyond static concept knowledge to better capture the representational dynamics of verbs and adverbs and their role in procedural processing. For instance, Kojima et al. [136] show that the observation of user behavior in a controlled environment can be used to improve model-generated instruction messages. Conversely, users adapt their instructions to a model based on the observed multimodal behavior [137].

5.3.1.2 Visual Saliency

Eye-tracking data can be used as a proxy for human attention in multimodal scenarios to learn how humans selectively attend to visual cues and seamlessly integrate them into a joint interpretation. For example, van Miltenburg et al. [138] recorded eye movements while participants generated a caption for an image that was presented on the screen. Takmaz et al. [139] added this eye-tracking information to the internal representation of a caption generation model based on the premise that speakers fixate salient areas in the image before mentioning them. These enriched representations led to more natural and more diverse image descriptions. The qualitative differences detected in this work reinforce the call for more fine-grained evaluation, as we discussed in Chap. 4. In a similar direction, Sood et al. [140] find that higher correlation between the attention of a multimodal model and human gaze patterns is a significant predictor of task performance in visual question answering indicating that models benefit from cognitively plausible representations. Dong et al. [141] show that integrating gaze information can also help in information retrieval to determine the relevance of documents, but that it is crucial to identify a suitable representational merging operation. It remains an open research question whether these task-specific findings generalize to a wider range of scenarios. In the following section, we discuss the main methods for cognitively grounding representations.

5.3.2   Cognitive Grounding

Under the umbrella of multimodal models, we mostly understand combinations of the textual and visual modality in natural language processing. Cognitively motivated approaches propose to interpret cognitive signals as an additional modality and repurpose the existing multimodal architectures. Whereas image processing and text processing models are well established individually, research on computational modeling of cognitive data is still in its early stages of development.

5.3.2.1 Signal Fusion

In early fusion approaches, representations for each modality are obtained separately and are directly concatenated to serve as input representations to a neural model. For example, augmenting language model representations of sentences with vector representations of eye-tracking features improves the performance on NLP tasks such as entity recognition [50], part-of-speech tagging [142], sentiment analysis [143], and the prediction of multiword expressions [144].

When we simply concatenate the input features in one joint decoder component, we leave it up to the model to identify the boundaries between the modalities and the different types of noise in the respective signals. Unfortunately, self-supervised representation learning [145] cannot yet be successfully applied due to the small sizes of datasets of cognitive signals related to language processing. It is therefore important to explicitly integrate knowledge about the strength and weaknesses of each cognitive signal type (see Chap. 3). For example, Bingel et al. [146] address the problem of low temporal resolution and hemodynamic delay of fMRI signals by using a Gaussian sliding window over the reading time when aligning words and signals. They show that the performance of a part-of-speech induction model is improved if the textual input is fused with this preprocessed cognitive signal. For EEG data, Hollenstein et al. [147] apply a Hilbert transformation on the time series data to combine the signals from multiple sensors. They then apply late fusion which means that the two modalities (text and EEG signal) correspond to two different input channels which can be realized as different network architectures (bidirectional LSTM vs convolutional inception in this case) and are only combined at the pre-final layer. The representations for each modality are jointly optimized with respect to a task (sentiment analysis and relation detection) but the boundaries between the modalities are clearly marked by the different channels.

5.3.2.2 Cognitively Informed Inference

The fusion methods come with the disadvantage that cognitive signals need to be available for every input unit, i.e., every token or sentence the model sees, whether at training or at test time. From an ethical standpoint, it is undesirable to expect cognitive signals during inference. This would mean that users need to consent to share their cognitive signals when using an application. Instead, cognitive signals should only be used for developing cognitively informed models that can generalize to inference scenarios.

The simplest approach to generalize cognitive signals is to perform type aggregation [50, 51]. This means that all cognitive features are averaged over the individual token occurrences. For example, if the token *turtle* appears three times in EEG recordings from reading, we average the signals of these three occurrences. As a result, a lexicon of word types with their averaged cognitive features is compiled. Tokens in the training data and in the test set are mapped to these features if the token occurs in the type-aggregated lexicon and to the unknown token otherwise. Unfortunately, this approach only works as a non-contextualized concept-based representation. Barrett et al. [51] apply type aggregation for part-of-speech induction and showed that the integration of eye-tracking data was beneficial even for ambiguous tokens that can occur in multiple part-of-speech classes.

In the methods described here, cognitive signals are used to enrich the input representation. Ren and Xiong [148] aim at linking text tokens and cognitive signals by using an attention mechanism that is supposed to disentangle text-specific information from the noise in the cognitive signal and facilitate token-level alignment (see Chap. 3). They combine EEG and eye-tracking signals and find that the cognitive information can be leveraged by the shared encoder for improved performance on several NLP tasks even if no cognitive signals are available during inference time. In Chap. 6, we further discuss how cognitive signals can be used to develop models with cognitively more plausible processing patterns, for example, by optimizing for an auxiliary cognitive task or by regulating the attention mechanism.

Augmenting models with cognitive signals requires comparing them to suitable baselines. On one hand, natural language understanding models are mostly optimized for performance on specific tasks and typically do not transfer well to other tasks or even other datasets of the same task [149]. On the other hand, cognitive processing signals are typically constrained to their experimental design and stimuli. These discrepancies may lead to limitations in the possible evaluation scenarios when leveraging cognitive signals to enhance language models [150]. Indeed, the improvements achieved with cognitive signals are often modest. Increasing the number of parameters on small training sets by adding cognitive features can make the models unstable [151]. It is therefore important to carefully compare the results to control tasks using robust baselines (see Sect. 5.1.2.3 and refraining from overpromising interpretations.

5.4   Ethical Aspects

Representational analyses usually only operate on intermediate computational representations and therefore have less direct ethical implications for humans. However, our methodological choices affect the conclusions we can draw from our experiments and our conclusions can have consequences on societal policies and the public perception of computational models. We discuss several methodological considerations that are relevant to the interpretation of representational analyses.

Earlier knowledge-based approaches for representing language were based on static databases and entity-relationship models that characterize the main concepts, their properties, and the relationships between them [4, 152]. In such knowledge bases, concepts can be enumerated and biases can be more explicitly quantified. High-dimensional contextualized representations better account for ambiguity and variation in language and can be adapted more dynamically. This ability to account for shades of grey comes with the downside of a lack of interpretability and transparency. When we want to analyze representational structure, we usually operate on simplifications of the representations. We project the data into fewer dimensions because our brain struggles with reasoning over high-dimensional relations. We build conceptual bridges by analyzing similarity relations and analogies, grouping concepts into clusters, or quantifying the probing success of binary properties. These analysis methods are important tools for better understanding representational structure, but we have to keep in mind that they are flawed simplifications of the representations.

If we are not transparent about the limitations of our methodological choices, we are more likely to fall into the trap of confirmation bias [153]. This bias describes our tendency to cherry-pick information that confirms our existing beliefs. In the scientific scenario, this is also known as experimenter bias and it affects both experimental design and result interpretation. If we do not force ourselves to set up proper control conditions, it is very likely that the experimental setup already favors a certain interpretation [154]. And if we are not transparent about negative results, we are tempted to selectively focus on experimental evidence that confirms our expectations [155]. We support the proposal by van Miltenburg et al. [156] to initiate a procedure of experimental pre-registration in natural language processing. Pre-registration requires transparently stating experimental hypotheses prior to obtaining results. When implemented meaningfully, this procedure could help to make the underlying assumptions of an experimental setup explicit and reduce the temptation to fall for post-hoc explanations.

Even if we pre-register our experiments, we might fail to identify the sources of empirical gains [157]. For example, when we enrich input representations with cognitive signals, we are effectively combining two signals that are both hard to interpret [109]. While this combination might lead to quantitative improvements, the gains could also simply be due to increasing the number of dimensions. In that case, adding cognitive information merely correlates with increased performance but the cause is the increased expressive power of the model. Permutation baselines can be an important control condition to check whether a task could be solved with similar performance under meaningless combinations of stimuli (see Sect. 5.1.2.3).

English is a morphologically shallow language which is reflected in the experimental setup of representational analyses. We cannot generalize approaches that are centered on words and subwords to languages with a non-concatenative morphology (for example, Semitic languages). The semantic organization of concepts varies between languages and is influenced by phylogenetic, geographical, and cultural factors [158–160]. Propagating static similarity labels across languages can be an oversimplified approach that is not empirically warranted due to cross-lingual semantic variations [32]. In the next chapter, we discuss how procedural analyses that explore how static representations are combined can reveal more information about cross-lingual differences.

References

1. Zellig S Harris. Distributional structure. *Word*, 10 (2-3): 146–162, 1954.
2. John R Firth. A synopsis of linguistic theory, 1930-1955. *Studies in linguistic analysis*, 1957.
3. Shimon Edelman. Representation is representation of similarities. *Behavioral and Brain Sciences*, 21 (4): 449–467, 1998. https://doi.org/10.1017/S0140525X98001253.

4. Christiane Fellbaum. Wordnet. In *Theory and applications of ontology: computer applications*, pages 231–243. Springer, 2010.

5. Pia Johanna Maria Sommerauer. *Diagnosing Semantic Properties in Distributional Representations of Word Meaning*. PhD thesis, Vrije Universiteit Amsterdam, June 2022.

6. Eneko Agirre, Enrique Alfonseca, Keith Hall, Jana Kravalova, Marius Paşca, and Aitor Soroa. A study on similarity and relatedness using distributional and WordNet-based approaches. In *Proceedings of Human Language Technologies: The 2009 Annual Conference of the North American Chapter of the Association for Computational Linguistics*, pages 19–27, Boulder, Colorado, June 2009. Association for Computational Linguistics. https://aclanthology.org/N09-1003.

7. Peter D Turney. Domain and function: A dual-space model of semantic relations and compositions. *Journal of artificial intelligence research*, 44: 533–585, 2012.

8. Felix Hill, Roi Reichart, and Anna Korhonen. Multi-modal models for concrete and abstract concept meaning. *Transactions of the Association for Computational Linguistics*, 2: 285–296, 2014.

9. Amos Tversky. Features of similarity. *Psychological review*, 84 (4): 327, 1977.

10. Kawin Ethayarajh. How contextual are contextualized word representations? Comparing the geometry of BERT, ELMo, and GPT-2 embeddings. In *Proceedings of the 2019 Conference on Empirical Methods in Natural Language Processing and the 9th International Joint Conference on Natural Language Processing (EMNLP-IJCNLP)*, pages 55–65, Hong Kong, China, November 2019. Association for Computational Linguistics. https://doi.org/10.18653/v1/D19-1006. https://aclanthology.org/D19-1006.

11. Nikos Athanasiou, Elias Iosif, and Alexandros Potamianos. Neural activation semantic models: Computational lexical semantic models of localized neural activations. In *Proceedings of the 27th International Conference on Computational Linguistics*, pages 2867–2878, Santa Fe, New Mexico, USA, August 2018. Association for Computational Linguistics. https://aclanthology.org/C18-1243.

12. Gino Brunner, Yang Liu, Damian Pascual, Oliver Richter, Massimiliano Ciaramita, and Roger Wattenhofer. On identifiability in transformers. In *International Conference on Learning Representations*, 2020. https://openreview.net/forum?id=BJg1f6EFDB.

13. William B. Dolan and Chris Brockett. Automatically constructing a corpus of sentential paraphrases. In *Proceedings of the Third International Workshop on Paraphrasing (IWP2005)*, 2005. https://aclanthology.org/I05-5002.

14. Eneko Agirre, Daniel Cer, Mona Diab, and Aitor Gonzalez-Agirre. SemEval-2012 task 6: A pilot on semantic textual similarity. In **SEM 2012: The First Joint Conference on Lexical and Computational Semantics—Volume 1: Proceedings of the main conference and the shared task, and Volume 2: Proceedings of the Sixth International Workshop on Semantic Evaluation (SemEval 2012)*, pages 385–393, Montréal, Canada, 7-8 June 2012. Association for Computational Linguistics. https://aclanthology.org/S12-1051.

15. Nils Reimers and Iryna Gurevych. Sentence-BERT: Sentence embeddings using Siamese BERT-networks. In *Proceedings of the 2019 Conference on Empirical Methods in Natural Language Processing and the 9th International Joint Conference on Natural Language Processing (EMNLP-IJCNLP)*, pages 3982–3992, Hong Kong, China, November 2019. Association for Computational Linguistics. https://doi.org/10.18653/v1/D19-1410. https://aclanthology.org/D19-1410.

16. Gregor Wiedemann, Steffen Remus, Avi Chawla, and Chris Biemann. Does bert make any sense? interpretable word sense disambiguation with contextualized embeddings. *arXiv preprint* arXiv:1909.10430, 2019.

17. Nora Hollenstein, Itziar Gonzalez-Dios, Lisa Beinborn, and Lena Jäger. Patterns of text read-
    ability in human and predicted eye movements. In *Proceedings of the Workshop on Cognitive
    Aspects of the Lexicon*, pages 1–15, Taipei, Taiwan, November 2022. Association for Compu-
    tational Linguistics. https://aclanthology.org/2022.cogalex-1.1.
18. David Mimno and Laure Thompson. The strange geometry of skip-gram with negative sampling.
    In *Proceedings of the 2017 Conference on Empirical Methods in Natural Language Processing*,
    pages 2873–2878, Copenhagen, Denmark, September 2017. Association for Computational
    Linguistics. https://doi.org/10.18653/v1/D17-1308. https://aclanthology.org/D17-1308.
19. Xingyu Cai, Jiaji Huang, Yuchen Bian, and Kenneth Church. Isotropy in the contextual embed-
    ding space: Clusters and manifolds. In *International Conference on Learning Representations*,
    2020.
20. Nicolas Papernot and Patrick McDaniel. Deep k-nearest neighbors: Towards confident, inter-
    pretable and robust deep learning. *arXiv preprint* arXiv:1803.04765, 2018.
21. Eric Wallace, Shi Feng, and Jordan Boyd-Graber. Interpreting neural networks with near-
    est neighbors. In *Proceedings of the 2018 EMNLP Workshop BlackboxNLP: Analyzing and
    Interpreting Neural Networks for NLP*, pages 136–144, Brussels, Belgium, November 2018.
    Association for Computational Linguistics. https://doi.org/10.18653/v1/W18-5416. https://
    aclanthology.org/W18-5416.
22. Been Kim, Martin Wattenberg, Justin Gilmer, Carrie Cai, James Wexler, Fernanda Viegas, et al.
    Interpretability beyond feature attribution: Quantitative testing with concept activation vectors
    (tcav). In *International conference on machine learning*, pages 2668–2677. PMLR, 2018.
23. Anders Søgaard. Neural speed reading audited. In *Findings of the Association for Computational
    Linguistics: EMNLP 2020*, pages 148–153, Online, November 2020. Association for Computa-
    tional Linguistics. https://doi.org/10.18653/v1/2020.findings-emnlp.14. https://aclanthology.
    org/2020.findings-emnlp.14.
24. Nils Reimers, Philip Beyer, and Iryna Gurevych. Task-oriented intrinsic evaluation of semantic
    textual similarity. In *Proceedings of COLING 2016, the 26th International Conference on
    Computational Linguistics: Technical Papers*, pages 87–96, Osaka, Japan, December 2016.
    The COLING 2016 Organizing Committee. https://aclanthology.org/C16-1009.
25. Felix Hill, Roi Reichart, and Anna Korhonen. SimLex-999: Evaluating semantic models with
    (genuine) similarity estimation. *Computational Linguistics*, 41 (4): 665–695, December 2015.
    https://doi.org/10.1162/COLI_a_00237. https://aclanthology.org/J15-4004.
26. Elia Bruni, Nam-Khanh Tran, and Marco Baroni. Multimodal distributional semantics. *Journal
    of artificial intelligence research*, 49: 1–47, 2014.
27. Lev Finkelstein, Evgeniy Gabrilovich, Yossi Matias, Ehud Rivlin, Zach Solan, Gadi Wolfman,
    and Eytan Ruppin. Placing search in context: The concept revisited. In *Proceedings of the 10th
    international conference on World Wide Web*, pages 406–414, 2001.
28. Minh-Thang Luong, Richard Socher, and Christopher D. Manning. Better word representations
    with recursive neural networks for morphology. In *CoNLL*, Sofia, Bulgaria, 2013.
29. David Jurgens, Saif Mohammad, Peter Turney, and Keith Holyoak. SemEval-2012 task 2: Mea-
    suring degrees of relational similarity. In **SEM 2012: The First Joint Conference on Lexical
    and Computational Semantics—Volume 1: Proceedings of the main conference and the shared
    task, and Volume 2: Proceedings of the Sixth International Workshop on Semantic Evaluation
    (SemEval 2012)*, pages 356–364, Montréal, Canada, 7-8 June 2012. Association for Computa-
    tional Linguistics. https://aclanthology.org/S12-1047.
30. Geoffrey Leech. Semantics. *Philosophy and Rhetoric*, 9 (1): 61–63, 1976.
31. Alexis Conneau and Douwe Kiela. SentEval: An evaluation toolkit for universal sentence rep-
    resentations. In *Proceedings of the Eleventh International Conference on Language Resources*

and Evaluation (LREC 2018), Miyazaki, Japan, May 2018. European Language Resources Association (ELRA). https://aclanthology.org/L18-1269.

32. Ella Rabinovich, Yang Xu, and Suzanne Stevenson. The typology of polysemy: A multilingual distributional framework. In *Proceedings of the 42nd Annual Meeting of the Cognitive Science Society*, pages 3370–3376, 2020.

33. Joseph Reisinger and Raymond Mooney. Cross-cutting models of lexical semantics. In *Proceedings of the 2011 Conference on Empirical Methods in Natural Language Processing*, pages 1405–1415, Edinburgh, Scotland, UK., July 2011. Association for Computational Linguistics. https://aclanthology.org/D11-1130.

34. Tomás Mikolov, Kai Chen, Greg Corrado, and Jeffrey Dean. Efficient estimation of word representations in vector space. In *1st International Conference on Learning Representations, ICLR 2013, Scottsdale, Arizona, USA, May 2-4, 2013, Workshop Track Proceedings*, 2013. http://arxiv.org/abs/1301.3781.

35. Peter D. Turney. Similarity of semantic relations. *Comput. Linguist.*, 32 (3): 379–416, sep 2006. ISSN 0891-2017. https://doi.org/10.1162/coli.2006.32.3.379.

36. Donald J Foss. A discourse on semantic priming. *Cognitive Psychology*, 14 (4): 590–607, 1982. ISSN 0010-0285. https://doi.org/10.1016/0010-0285(82)90020-2. https://www.sciencedirect.com/science/article/pii/0010028582900202.

37. Max Coltheart. The mrc psycholinguistic database. *The Quarterly Journal of Experimental Psychology Section A*, 33 (4): 497–505, 1981.

38. Anna Rogers, Aleksandr Drozd, and Bofang Li. The (too many) problems of analogical reasoning with word vectors. In *Proceedings of the 6th Joint Conference on Lexical and Computational Semantics (*SEM 2017)*, pages 135–148, Vancouver, Canada, August 2017. Association for Computational Linguistics. https://doi.org/10.18653/v1/S17-1017. https://aclanthology.org/S17-1017.

39. Allyson Ettinger and Tal Linzen. Evaluating vector space models using human semantic priming results. In *Proceedings of the 1st Workshop on Evaluating Vector-Space Representations for NLP*, pages 72–77, Berlin, Germany, August 2016. Association for Computational Linguistics. https://doi.org/10.18653/v1/W16-2513. https://aclanthology.org/W16-2513.

40. Keith A Hutchison, David A Balota, James H Neely, Michael J Cortese, Emily R Cohen-Shikora, Chi-Shing Tse, Melvin J Yap, Jesse J Bengson, Dale Niemeyer, and Erin Buchanan. The semantic priming project. *Behavior research methods*, 45 (4): 1099–1114, 2013.

41. Jeremy Auguste, Arnaud Rey, and Benoit Favre. Evaluation of word embeddings against cognitive processes: primed reaction times in lexical decision and naming tasks. In *Proceedings of the 2nd Workshop on Evaluating Vector Space Representations for NLP*, pages 21–26, Copenhagen, Denmark, September 2017. Association for Computational Linguistics. https://doi.org/10.18653/v1/W17-5304. https://aclanthology.org/W17-5304.

42. Nikolaus Kriegeskorte, Marieke Mur, and Peter A Bandettini. Representational similarity analysis-connecting the branches of systems neuroscience. *Frontiers in systems neuroscience*, 2: 4, 2008.

43. Samira Abnar, Lisa Beinborn, Rochelle Choenni, and Willem Zuidema. Blackbox meets blackbox: Representational similarity & stability analysis of neural language models and brains. In *Proceedings of the 2019 ACL Workshop BlackboxNLP: Analyzing and Interpreting Neural Networks for NLP*, pages 191–203, Florence, Italy, August 2019. Association for Computational Linguistics. https://doi.org/10.18653/v1/W19-4820. https://aclanthology.org/W19-4820.

44. Mostafa Abdou, Artur Kulmizev, Felix Hill, Daniel M. Low, and Anders Søgaard. Higher-order comparisons of sentence encoder representations. In *Proceedings of the 2019 Conference on Empirical Methods in Natural Language Processing and the 9th International Joint Conference on Natural Language Processing (EMNLP-IJCNLP)*, pages 5838–5845, Hong Kong, China,

November 2019. Association for Computational Linguistics. https://doi.org/10.18653/v1/D19-1593. https://aclanthology.org/D19-1593.

45. Grzegorz Chrupała and Afra Alishahi. Correlating neural and symbolic representations of language. In *Proceedings of the 57th Annual Meeting of the Association for Computational Linguistics*, pages 2952–2962, Florence, Italy, July 2019. Association for Computational Linguistics. https://doi.org/10.18653/v1/P19-1283. https://aclanthology.org/P19-1283.

46. Anders Søgaard. Explainable natural language processing. *Synthesis Lectures on Human Language Technologies*, 14 (3): 1–123, 2021.

47. Alexander Craik, Yongtian He, and Jose L Contreras-Vidal. Deep learning for electroencephalogram (eeg) classification tasks: a review. *Journal of neural engineering*, 16 (3): 031001, 2019.

48. Rotem Dror, Gili Baumer, Segev Shlomov, and Roi Reichart. The hitchhiker's guide to testing statistical significance in natural language processing. In *Proceedings of the 56th Annual Meeting of the Association for Computational Linguistics (Volume 1: Long Papers)*, pages 1383–1392, Melbourne, Australia, July 2018. Association for Computational Linguistics. https://doi.org/10.18653/v1/P18-1128. https://aclanthology.org/P18-1128.

49. Yang Liu, Alan Medlar, and Dorota Glowacka. Statistically significant detection of semantic shifts using contextual word embeddings. In *Proceedings of the 2nd Workshop on Evaluation and Comparison of NLP Systems*, pages 104–113, Punta Cana, Dominican Republic, November 2021b. Association for Computational Linguistics. https://doi.org/10.18653/v1/2021.eval4nlp-1.11. https://aclanthology.org/2021.eval4nlp-1.11.

50. Nora Hollenstein and Ce Zhang. Entity recognition at first sight: Improving NER with eye movement information. In *Proceedings of the 2019 Conference of the North American Chapter of the Association for Computational Linguistics: Human Language Technologies, Volume 1 (Long and Short Papers)*, pages 1–10, Minneapolis, Minnesota, June 2019. Association for Computational Linguistics. https://doi.org/10.18653/v1/N19-1001. https://aclanthology.org/N19-1001.

51. Maria Barrett, Joachim Bingel, Frank Keller, and Anders Søgaard. Weakly supervised part-of-speech tagging using eye-tracking data. In *Proceedings of the 54th Annual Meeting of the Association for Computational Linguistics (Volume 2: Short Papers)*, pages 579–584, Berlin, Germany, August 2016. Association for Computational Linguistics. https://doi.org/10.18653/v1/P16-2094. https://aclanthology.org/P16-2094.

52. Nora Hollenstein and Lisa Beinborn. Relative importance in sentence processing. In *Proceedings of the 59th Annual Meeting of the Association for Computational Linguistics and the 11th International Joint Conference on Natural Language Processing (Volume 2: Short Papers)*, pages 141–150, Online, August 2021. Association for Computational Linguistics. https://doi.org/10.18653/v1/2021.acl-short.19. https://aclanthology.org/2021.acl-short.19.

53. G Rupert Jr et al. *Simultaneous statistical inference*. Springer Science & Business Media, 2012.

54. Carlo Bonferroni. Teoria statistica delle classi e calcolo delle probabilita. *Pubblicazioni del R Istituto Superiore di Scienze Economiche e Commericiali di Firenze*, 8: 3–62, 1936.

55. Rotem Dror, Lotem Peled-Cohen, Segev Shlomov, and Roi Reichart. Statistical significance testing for natural language processing. *Synthesis Lectures on Human Language Technologies*, 13 (2): 1–116, 2020. https://doi.org/10.2200/S00994ED1V01Y202002HLT045. https://doi.org/10.2200/S00994ED1V01Y202002HLT045.

56. Yoav Benjamini and Yosef Hochberg. Controlling the false discovery rate: a practical and powerful approach to multiple testing. *Journal of the Royal statistical society: series B (Methodological)*, 57 (1): 289–300, 1995.

57. Chandler May, Alex Wang, Shikha Bordia, Samuel R. Bowman, and Rachel Rudinger. On measuring social biases in sentence encoders. In *Proceedings of the 2019 Conference of the North American Chapter of the Association for Computational Linguistics: Human Language Technologies, Volume 1 (Long and Short Papers)*, pages 622–628, Minneapolis, Minnesota,

June 2019. Association for Computational Linguistics. https://doi.org/10.18653/v1/N19-1063. https://aclanthology.org/N19-1063.

58. William Timkey and Marten van Schijndel. All bark and no bite: Rogue dimensions in transformer language models obscure representational quality. In *Proceedings of the 2021 Conference on Empirical Methods in Natural Language Processing*, pages 4527–4546, Online and Punta Cana, Dominican Republic, November 2021. Association for Computational Linguistics. https://doi.org/10.18653/v1/2021.emnlp-main.372. https://aclanthology.org/2021.emnlp-main.372.

59. Kaitlyn Zhou, Kawin Ethayarajh, Dallas Card, and Dan Jurafsky. Problems with cosine as a measure of embedding similarity for high frequency words. In *Proceedings of the 60th Annual Meeting of the Association for Computational Linguistics (Volume 2: Short Papers)*, pages 401–423, Dublin, Ireland, May 2022. Association for Computational Linguistics. https://doi.org/10.18653/v1/2022.acl-short.45. https://aclanthology.org/2022.acl-short.45.

60. Allyson Ettinger, Ahmed Elgohary, and Philip Resnik. Probing for semantic evidence of composition by means of simple classification tasks. In *Proceedings of the 1st Workshop on Evaluating Vector-Space Representations for NLP*, pages 134–139, Berlin, Germany, August 2016. Association for Computational Linguistics. https://doi.org/10.18653/v1/W16-2524. https://aclanthology.org/W16-2524.

61. Kristina Gulordava, Piotr Bojanowski, Edouard Grave, Tal Linzen, and Marco Baroni. Colorless green recurrent networks dream hierarchically. In *Proceedings of the 2018 Conference of the North American Chapter of the Association for Computational Linguistics: Human Language Technologies, Volume 1 (Long Papers)*, pages 1195–1205, New Orleans, Louisiana, June 2018. Association for Computational Linguistics. https://doi.org/10.18653/v1/N18-1108. https://aclanthology.org/N18-1108.

62. Mario Giulianelli, Jack Harding, Florian Mohnert, Dieuwke Hupkes, and Willem Zuidema. Under the hood: Using diagnostic classifiers to investigate and improve how language models track agreement information. In *Proceedings of the 2018 EMNLP Workshop BlackboxNLP: Analyzing and Interpreting Neural Networks for NLP*, pages 240–248, Brussels, Belgium, November 2018. Association for Computational Linguistics. https://doi.org/10.18653/v1/W18-5426. https://aclanthology.org/W18-5426.

63. Christopher D Manning, Kevin Clark, John Hewitt, Urvashi Khandelwal, and Omer Levy. Emergent linguistic structure in artificial neural networks trained by self-supervision. *Proceedings of the National Academy of Sciences*, 117 (48): 30046–30054, 2020.

64. Nadir Durrani, Hassan Sajjad, and Fahim Dalvi. How transfer learning impacts linguistic knowledge in deep NLP models? In *Findings of the Association for Computational Linguistics: ACL-IJCNLP 2021*, pages 4947–4957, Online, August 2021. Association for Computational Linguistics. https://doi.org/10.18653/v1/2021.findings-acl.438. https://aclanthology.org/2021.findings-acl.438.

65. Ionut-Teodor Sorodoc, Kristina Gulordava, and Gemma Boleda. Probing for referential information in language models. In *Proceedings of the 58th Annual Meeting of the Association for Computational Linguistics*, pages 4177–4189, Online, July 2020. Association for Computational Linguistics. https://doi.org/10.18653/v1/2020.acl-main.384. https://aclanthology.org/2020.acl-main.384.

66. Richard Futrell, Ethan Wilcox, Takashi Morita, Peng Qian, Miguel Ballesteros, and Roger Levy. Neural language models as psycholinguistic subjects: Representations of syntactic state. In *Proceedings of the 2019 Conference of the North American Chapter of the Association for Computational Linguistics: Human Language Technologies, Volume 1 (Long and Short Papers)*, pages 32–42, Minneapolis, Minnesota, June 2019. Association for Computational Linguistics. https://doi.org/10.18653/v1/N19-1004. https://aclanthology.org/N19-1004.

67. Emily Reif, Ann Yuan, Martin Wattenberg, Fernanda B Viegas, Andy Coenen, Adam Pearce, and Been Kim. Visualizing and measuring the geometry of bert. *Advances in Neural Information Processing Systems*, 32, 2019.

68. Alex Warstadt, Yu Cao, Ioana Grosu, Wei Peng, Hagen Blix, Yining Nie, Anna Alsop, Shikha Bordia, Haokun Liu, Alicia Parrish, Sheng-Fu Wang, Jason Phang, Anhad Mohananey, Phu Mon Htut, Paloma Jeretic, and Samuel R. Bowman. Investigating BERT's knowledge of language: Five analysis methods with NPIs. In *Proceedings of the 2019 Conference on Empirical Methods in Natural Language Processing and the 9th International Joint Conference on Natural Language Processing (EMNLP-IJCNLP)*, pages 2877–2887, Hong Kong, China, November 2019. Association for Computational Linguistics. https://doi.org/10.18653/v1/D19-1286. https://aclanthology.org/D19-1286.

69. Anna Rogers, Olga Kovaleva, and Anna Rumshisky. A primer in BERTology: What we know about how BERT works. *Transactions of the Association for Computational Linguistics*, 8: 842–866, 2020. https://doi.org/10.1162/tacl_a_00349. https://aclanthology.org/2020.tacl-1.54.

70. Ian Tenney, Patrick Xia, Berlin Chen, Alex Wang, Adam Poliak, R Thomas McCoy, Najoung Kim, Benjamin Van Durme, Samuel R Bowman, Dipanjan Das, et al. What do you learn from context? probing for sentence structure in contextualized word representations. In *International Conference on Learning Representations*, 2018.

71. Alexis Conneau, German Kruszewski, Guillaume Lample, Loïc Barrault, and Marco Baroni. What you can cram into a single $&!#* vector: Probing sentence embeddings for linguistic properties. In *Proceedings of the 56th Annual Meeting of the Association for Computational Linguistics (Volume 1: Long Papers)*, pages 2126–2136, Melbourne, Australia, July 2018. Association for Computational Linguistics. https://doi.org/10.18653/v1/P18-1198. https://aclanthology.org/P18-1198.

72. Robyn Speer, Joshua Chin, and Catherine Havasi. Conceptnet 5.5: an open multilingual graph of general knowledge. In *Proceedings of the Thirty-First AAAI Conference on Artificial Intelligence*, pages 4444–4451, 2017.

73. Zied Bouraoui, Jose Camacho-Collados, and Steven Schockaert. Inducing relational knowledge from bert. In *Proceedings of the AAAI Conference on Artificial Intelligence*, volume 34, pages 7456–7463, 2020.

74. Fabio Petroni, Tim Rocktäschel, Sebastian Riedel, Patrick Lewis, Anton Bakhtin, Yuxiang Wu, and Alexander Miller. Language models as knowledge bases? In *Proceedings of the 2019 Conference on Empirical Methods in Natural Language Processing and the 9th International Joint Conference on Natural Language Processing (EMNLP-IJCNLP)*, pages 2463–2473, Hong Kong, China, November 2019. Association for Computational Linguistics. https://doi.org/10.18653/v1/D19-1250. https://aclanthology.org/D19-1250.

75. Bill Yuchen Lin, Seyeon Lee, Rahul Khanna, and Xiang Ren. Birds have four legs?! NumerSense: Probing Numerical Commonsense Knowledge of Pre-Trained Language Models. In *Proceedings of the 2020 Conference on Empirical Methods in Natural Language Processing (EMNLP)*, pages 6862–6868, Online, November 2020. Association for Computational Linguistics. https://doi.org/10.18653/v1/2020.emnlp-main.557. https://aclanthology.org/2020.emnlp-main.557.

76. Eric Mitchell, Joseph Noh, Siyan Li, Will Armstrong, Ananth Agarwal, Patrick Liu, Chelsea Finn, and Christopher Manning. Enhancing self-consistency and performance of pre-trained language models through natural language inference. In *Proceedings of the 2022 Conference on Empirical Methods in Natural Language Processing*, pages 1754–1768, Abu Dhabi, United Arab Emirates, December 2022. Association for Computational Linguistics. https://aclanthology.org/2022.emnlp-main.115.

77. Yonatan Belinkov. Probing classifiers: Promises, shortcomings, and advances. *Computational Linguistics*, 48 (1): 207–219, March 2022. https://doi.org/10.1162/coli_a_00422. https://aclanthology.org/2022.cl-1.7.

78. Dieuwke Hupkes, Sara Veldhoen, and Willem Zuidema. Visualisation and 'diagnostic classifiers' reveal how recurrent and recursive neural networks process hierarchical structure. *Journal of Artificial Intelligence Research*, 61: 907–926, 2018.

79. John Hewitt and Percy Liang. Designing and interpreting probes with control tasks. In *Proceedings of the 2019 Conference on Empirical Methods in Natural Language Processing and the 9th International Joint Conference on Natural Language Processing (EMNLP-IJCNLP)*, pages 2733–2743, Hong Kong, China, November 2019. Association for Computational Linguistics. https://doi.org/10.18653/v1/D19-1275. https://aclanthology.org/D19-1275.

80. Tiago Pimentel, Josef Valvoda, Rowan Hall Maudslay, Ran Zmigrod, Adina Williams, and Ryan Cotterell. Information-theoretic probing for linguistic structure. In *Proceedings of the 58th Annual Meeting of the Association for Computational Linguistics*, pages 4609–4622, Online, July 2020. Association for Computational Linguistics. https://doi.org/10.18653/v1/2020.acl-main.420. https://aclanthology.org/2020.acl-main.420.

81. Lucas Torroba Hennigen, Adina Williams, and Ryan Cotterell. Intrinsic probing through dimension selection. In *Proceedings of the 2020 Conference on Empirical Methods in Natural Language Processing (EMNLP)*, pages 197–216, Online, November 2020. Association for Computational Linguistics. https://doi.org/10.18653/v1/2020.emnlp-main.15. https://aclanthology.org/2020.emnlp-main.15.

82. Yair Lakretz, German Kruszewski, Theo Desbordes, Dieuwke Hupkes, Stanislas Dehaene, and Marco Baroni. The emergence of number and syntax units in LSTM language models. In *Proceedings of the 2019 Conference of the North American Chapter of the Association for Computational Linguistics: Human Language Technologies, Volume 1 (Long and Short Papers)*, pages 11–20, Minneapolis, Minnesota, June 2019. Association for Computational Linguistics. https://doi.org/10.18653/v1/N19-1002. https://aclanthology.org/N19-1002.

83. Jiwei Li, Xinlei Chen, Eduard Hovy, and Dan Jurafsky. Visualizing and understanding neural models in NLP. In *Proceedings of the 2016 Conference of the North American Chapter of the Association for Computational Linguistics: Human Language Technologies*, pages 681–691, San Diego, California, June 2016. Association for Computational Linguistics. https://doi.org/10.18653/v1/N16-1082. https://aclanthology.org/N16-1082.

84. Hassan Sajjad, Nadir Durrani, and Fahim Dalvi. Neuron-level interpretation of deep nlp models: A survey. *Transactions of the Association for Computational Linguistics*, 10: 1285–1303, 2022.

85. Yanai Elazar, Shauli Ravfogel, Alon Jacovi, and Yoav Goldberg. Amnesic probing: Behavioral explanation with amnesic counterfactuals. *Transactions of the Association for Computational Linguistics*, 9: 160–175, 2021. https://doi.org/10.1162/tacl_a_00359. https://aclanthology.org/2021.tacl-1.10.

86. Shauli Ravfogel, Yanai Elazar, Hila Gonen, Michael Twiton, and Yoav Goldberg. Null it out: Guarding protected attributes by iterative nullspace projection. In *Proceedings of the 58th Annual Meeting of the Association for Computational Linguistics*, pages 7237–7256, Online, July 2020. Association for Computational Linguistics. https://doi.org/10.18653/v1/2020.acl-main.647. https://aclanthology.org/2020.acl-main.647.

87. Thomas Naselaris, Kendrick N Kay, Shinji Nishimoto, and Jack L Gallant. Encoding and decoding in fmri. *Neuroimage*, 56 (2): 400–410, 2011.

88. Alexander G Huth, Wendy A De Heer, Thomas L Griffiths, Frédéric E Theunissen, and Jack L Gallant. Natural speech reveals the semantic maps that tile human cerebral cortex. *Nature*, 532 (7600): 453–458, 2016.

89. Brian Murphy, Leila Wehbe, and Alona Fyshe. Decoding language from the brain. *Language, cognition, and computational models*, pages 53–80, 2018.

90. Jon Gauthier and Anna Ivanova. Does the brain represent words? an evaluation of brain decoding studies of language understanding. *arXiv preprint* arXiv:1806.00591, 2018.

91. Damian Pascual, Béni Egressy, Nicolas Affolter, Yiming Cai, Oliver Richter, and Roger Wattenhofer. Improving brain decoding methods and evaluation. In *ICASSP 2022-2022 IEEE International Conference on Acoustics, Speech and Signal Processing (ICASSP)*, pages 1476–1480. IEEE, 2022.

92. Tom M Mitchell, Svetlana V Shinkareva, Andrew Carlson, Kai-Min Chang, Vicente L Malave, Robert A Mason, and Marcel Adam Just. Predicting human brain activity associated with the meanings of nouns. *science*, 320 (5880): 1191–1195, 2008.

93. Samira Abnar, Rasyan Ahmed, Max Mijnheer, and Willem Zuidema. Experiential, distributional and dependency-based word embeddings have complementary roles in decoding brain activity. In *Proceedings of the 8th Workshop on Cognitive Modeling and Computational Linguistics (CMCL 2018)*, pages 57–66, Salt Lake City, Utah, January 2018. Association for Computational Linguistics. https://doi.org/10.18653/v1/W18-0107. https://aclanthology.org/W18-0107.

94. João António Rodrigues, Ruben Branco, João Silva, Chakaveh Saedi, and António Branco. Predicting brain activation with WordNet embeddings. In *Proceedings of the Eight Workshop on Cognitive Aspects of Computational Language Learning and Processing*, pages 1–5, Melbourne, July 2018. Association for Computational Linguistics. https://doi.org/10.18653/v1/W18-2801. https://aclanthology.org/W18-2801.

95. Haoyan Xu, Brian Murphy, and Alona Fyshe. BrainBench: A brain-image test suite for distributional semantic models. In *Proceedings of the 2016 Conference on Empirical Methods in Natural Language Processing*, pages 2017–2021, Austin, Texas, November 2016. Association for Computational Linguistics. https://doi.org/10.18653/v1/D16-1213. https://aclanthology.org/D16-1213.

96. Anders Søgaard. Evaluating word embeddings with fMRI and eye-tracking. In *Proceedings of the 1st Workshop on Evaluating Vector-Space Representations for NLP*, pages 116–121, Berlin, Germany, August 2016. Association for Computational Linguistics. https://doi.org/10.18653/v1/W16-2521. https://aclanthology.org/W16-2521.

97. Nora Hollenstein, Antonio de la Torre, Nicolas Langer, and Ce Zhang. CogniVal: A framework for cognitive word embedding evaluation. In *Proceedings of the 23rd Conference on Computational Natural Language Learning (CoNLL)*, pages 538–549, Hong Kong, China, November 2019b. Association for Computational Linguistics. https://doi.org/10.18653/v1/K19-1050. https://aclanthology.org/K19-1050.

98. Alona Fyshe, Partha P. Talukdar, Brian Murphy, and Tom M. Mitchell. Interpretable semantic vectors from a joint model of brain- and text- based meaning. In *Proceedings of the 52nd Annual Meeting of the Association for Computational Linguistics (Volume 1: Long Papers)*, pages 489–499, Baltimore, Maryland, June 2014. Association for Computational Linguistics. https://doi.org/10.3115/v1/P14-1046. https://aclanthology.org/P14-1046.

99. Angeliki Lazaridou, Nghia The Pham, and Marco Baroni. Combining language and vision with a multimodal skip-gram model. In *Proceedings of the 2015 Conference of the North American Chapter of the Association for Computational Linguistics: Human Language Technologies*, pages 153–163, Denver, Colorado, May–June 2015. Association for Computational Linguistics. https://doi.org/10.3115/v1/N15-1016. https://aclanthology.org/N15-1016.

100. Lisa Beinborn, Teresa Botschen, and Iryna Gurevych. Multimodal grounding for language processing. In *Proceedings of the 27th International Conference on Computational Linguistics*, pages 2325–2339, Santa Fe, New Mexico, USA, August 2018. Association for Computational Linguistics. https://aclanthology.org/C18-1197.

101. Luana Bulat, Stephen Clark, and Ekaterina Shutova. Speaking, seeing, understanding: Correlating semantic models with conceptual representation in the brain. In *Proceedings of the 2017 Conference on Empirical Methods in Natural Language Processing*, pages 1081–1091, Copenhagen, Denmark, September 2017. Association for Computational Linguistics. https://doi.org/10.18653/v1/D17-1113. https://aclanthology.org/D17-1113.

102. Andrew J Anderson, Douwe Kiela, Stephen Clark, and Massimo Poesio. Visually grounded and textual semantic models differentially decode brain activity associated with concrete and abstract nouns. *Transactions of the Association for Computational Linguistics*, 5: 17–30, 2017.

103. Eva Hendrikx and Lisa Beinborn. The fluidity of concept representations in human brain signals. *arXiv preprint* arXiv:2002.08880, 2020.

104. Liberty S Hamilton and Alexander G Huth. The revolution will not be controlled: natural stimuli in speech neuroscience. *Language, cognition and neuroscience*, 35 (5): 573–582, 2020.

105. Francisco Pereira, Bin Lou, Brianna Pritchett, Samuel Ritter, Samuel J. Gershman, Nancy Kanwisher, Matthew Botvinick, and Evelina Fedorenko. Toward a universal decoder of linguistic meaning from brain activation. *Nature communications*, 9: 963, 2018. https://www.nature.com/articles/s41467-018-03068-4.pdf.

106. Leila Wehbe, Ashish Vaswani, Kevin Knight, and Tom Mitchell. Aligning context-based statistical models of language with brain activity during reading. In *Proceedings of the 2014 Conference on Empirical Methods in Natural Language Processing (EMNLP)*, pages 233–243, Doha, Qatar, October 2014. Association for Computational Linguistics. https://doi.org/10.3115/v1/D14-1030. https://aclanthology.org/D14-1030.

107. Morteza Dehghani, Reihane Boghrati, Kingson Man, Joe Hoover, Sarah I Gimbel, Ashish Vaswani, Jason D Zevin, Mary Helen Immordino-Yang, Andrew S Gordon, Antonio Damasio, et al. Decoding the neural representation of story meanings across languages. *Human brain mapping*, 38 (12): 6096–6106, 2017.

108. Lisa Beinborn, Samira Abnar, and Rochelle Choenni. Robust evaluation of language-brain encoding experiments. *International Journal of Computational Linguistics and Applications*, 2019.

109. Mariya Toneva and Leila Wehbe. Interpreting and improving natural-language processing (in machines) with natural language-processing (in the brain). *Advances in Neural Information Processing Systems*, 32, 2019.

110. Martin Schrimpf, Idan Asher Blank, Greta Tuckute, Carina Kauf, Eghbal A Hosseini, Nancy Kanwisher, Joshua B Tenenbaum, and Evelina Fedorenko. The neural architecture of language: Integrative modeling converges on predictive processing. *Proceedings of the National Academy of Sciences*, 118 (45), 2021.

111. Jonathan R Brennan, Edward P Stabler, Sarah E Van Wagenen, Wen-Ming Luh, and John T Hale. Abstract linguistic structure correlates with temporal activity during naturalistic comprehension. *Brain and language*, 157: 81–94, 2016.

112. Jonathan R Brennan and John T Hale. Hierarchical structure guides rapid linguistic predictions during naturalistic listening. *PloS one*, 14 (1): e0207741, 2019.

113. Ariel Goldstein, Zaid Zada, Eliav Buchnik, Mariano Schain, Amy Price, Bobbi Aubrey, Samuel A Nastase, Amir Feder, Dotan Emanuel, Alon Cohen, et al. Shared computational principles for language processing in humans and deep language models. *Nature neuroscience*, 25 (3): 369–380, 2022.

114. Alessandro Lopopolo, Stefan L Frank, Antal Van den Bosch, Annabel Nijhof, and Roel M Willems. The narrative brain dataset (nbd), an fMRI dataset for the study of natural language processing in the brain. In *LREC 2018 Workshop" Linguistic and Neuro-Cognitive Resources (LiNCR)*, pages 8–11. LREC, 2018. http://lrec-conf.org/workshops/lrec2018/W9/pdf/1_W9.pdf.

115. Shailee Jain and Alexander Huth. Incorporating context into language encoding models for fmri. *Advances in neural information processing systems*, 31, 2018.
116. Charlotte Caucheteux and Jean-Rémi King. Brains and algorithms partially converge in natural language processing. *Communications biology*, 5 (1): 1–10, 2022.
117. Lawrence W Barsalou et al. Grounded cognition. *Annual review of psychology*, 59 (1): 617–645, 2008.
118. Alvin I Goldman et al. *Simulating minds: The philosophy, psychology, and neuroscience of mindreading*. Oxford University Press on Demand, 2006.
119. Benjamin Bergen. Embodiment. *Cognitive Linguistics: Foundations of language*, pages 11–35, 2019.
120. Felix Hill, Roi Reichart, and Anna Korhonen. Multi-modal models for concrete and abstract concept meaning. *Transactions of the Association for Computational Linguistics*, 2: 285–296, 2014. https://doi.org/10.1162/tacl_a_00183. https://aclanthology.org/Q14-1023.
121. Carina Silberer and Mirella Lapata. Grounded models of semantic representation. In *Proceedings of the 2012 Joint Conference on Empirical Methods in Natural Language Processing and Computational Natural Language Learning*, pages 1423–1433, Jeju Island, Korea, July 2012. Association for Computational Linguistics. https://aclanthology.org/D12-1130.
122. Elia Bruni, Gemma Boleda, Marco Baroni, and Nam-Khanh Tran. Distributional semantics in technicolor. In *Proceedings of the 50th Annual Meeting of the Association for Computational Linguistics (Volume 1: Long Papers)*, pages 136–145, Jeju Island, Korea, July 2012. Association for Computational Linguistics. https://aclanthology.org/P12-1015.
123. Marco Baroni. Grounding distributional semantics in the visual world. *Language and Linguistics Compass*, 10 (1): 3–13, 2016.
124. Nasrin Mostafazadeh, Chris Brockett, Bill Dolan, Michel Galley, Jianfeng Gao, Georgios Spithourakis, and Lucy Vanderwende. Image-grounded conversations: Multimodal context for natural question and response generation. In *Proceedings of the Eighth International Joint Conference on Natural Language Processing (Volume 1: Long Papers)*, pages 462–472, Taipei, Taiwan, November 2017. Asian Federation of Natural Language Processing. https://aclanthology.org/I17-1047.
125. Ravi Shekhar, Ece Takmaz, Raquel Fernández, and Raffaella Bernardi. Evaluating the representational hub of language and vision models. In *Proceedings of the 13th International Conference on Computational Semantics–Long Papers*, pages 211–222, Gothenburg, Sweden, May 2019. Association for Computational Linguistics. https://doi.org/10.18653/v1/W19-0418. https://aclanthology.org/W19-0418.
126. Jing Han, Zixing Zhang, Zhao Ren, and Björn Schuller. Emobed: Strengthening monomodal emotion recognition via training with crossmodal emotion embeddings. *IEEE Transactions on Affective Computing*, 12 (3): 553–564, 2019.
127. Tadas Baltrušaitis, Chaitanya Ahuja, and Louis-Philippe Morency. Multimodal machine learning: A survey and taxonomy. *IEEE transactions on pattern analysis and machine intelligence*, 41 (2): 423–443, 2018.
128. Yen-Chun Chen, Linjie Li, Licheng Yu, Ahmed El Kholy, Faisal Ahmed, Zhe Gan, Yu Cheng, and Jingjing Liu. Uniter: Universal image-text representation learning. In *European conference on computer vision*, pages 104–120. Springer, 2020.
129. Gen Li, Nan Duan, Yuejian Fang, Ming Gong, and Daxin Jiang. Unicoder-vl: A universal encoder for vision and language by cross-modal pre-training. In *Proceedings of the AAAI Conference on Artificial Intelligence*, volume 34, pages 11336–11344, 2020.
130. Weijie Su, Xizhou Zhu, Yue Cao, Bin Li, Lewei Lu, Furu Wei, and Jifeng Dai. VL-BERT: Pre-training of generic visual-linguistic representations. *arXiv preprint* arXiv:1908.08530, 2019.

131. Hao Tan and Mohit Bansal. LXMERT: Learning cross-modality encoder representations from transformers. In *Proceedings of the 2019 Conference on Empirical Methods in Natural Language Processing and the 9th International Joint Conference on Natural Language Processing (EMNLP-IJCNLP)*, pages 5100–5111, Hong Kong, China, November 2019. Association for Computational Linguistics. https://doi.org/10.18653/v1/D19-1514. https://aclanthology.org/D19-1514.

132. Sandro Pezzelle, Claudio Greco, Greta Gandolfi, Eleonora Gualdoni, and Raffaella Bernardi. Be Different to Be Better! A Benchmark to Leverage the Complementarity of Language and Vision. In *Findings of the Association for Computational Linguistics: EMNLP 2020*, pages 2751–2767, Online, November 2020. Association for Computational Linguistics. https://doi.org/10.18653/v1/2020.findings-emnlp.248. https://aclanthology.org/2020.findings-emnlp.248.

133. Adam Dahlgren Lindström, Johanna Björklund, Suna Bensch, and Frank Drewes. Probing multimodal embeddings for linguistic properties: the visual-semantic case. In *Proceedings of the 28th International Conference on Computational Linguistics*, pages 730–744, Barcelona, Spain (Online), December 2020. International Committee on Computational Linguistics. https://doi.org/10.18653/v1/2020.coling-main.64. https://aclanthology.org/2020.coling-main.64.

134. Emmanuelle Salin, Badreddine Farah, Stéphane Ayache, and Benoit Favre. Are vision-language transformers learning multimodal representations? a probing perspective. In *AAAI 2022*, 2022.

135. Alessandro Suglia, Yonatan Bisk, Ioannis Konstas, Antonio Vergari, Emanuele Bastianelli, Andrea Vanzo, and Oliver Lemon. An empirical study on the generalization power of neural representations learned via visual guessing games. In *Proceedings of the 16th Conference of the European Chapter of the Association for Computational Linguistics: Main Volume*, pages 2135–2144, Online, April 2021. Association for Computational Linguistics. https://doi.org/10.18653/v1/2021.eacl-main.183. https://aclanthology.org/2021.eacl-main.183.

136. Noriyuki Kojima, Alane Suhr, and Yoav Artzi. Continual learning for grounded instruction generation by observing human following behavior. *Transactions of the Association for Computational Linguistics*, 9: 1303–1319, 2021. https://doi.org/10.1162/tacl_a_00428. https://aclanthology.org/2021.tacl-1.77.

137. Alane Suhr, Claudia Yan, Jack Schluger, Stanley Yu, Hadi Khader, Marwa Mouallem, Iris Zhang, and Yoav Artzi. Executing instructions in situated collaborative interactions. In *Proceedings of the 2019 Conference on Empirical Methods in Natural Language Processing and the 9th International Joint Conference on Natural Language Processing (EMNLP-IJCNLP)*, pages 2119–2130, Hong Kong, China, November 2019. Association for Computational Linguistics. https://doi.org/10.18653/v1/D19-1218. https://aclanthology.org/D19-1218.

138. Emiel van Miltenburg, Ákos Kádár, Ruud Koolen, and Emiel Krahmer. DIDEC: The Dutch image description and eye-tracking corpus. In *Proceedings of the 27th International Conference on Computational Linguistics*, pages 3658–3669, Santa Fe, New Mexico, USA, August 2018. Association for Computational Linguistics. https://aclanthology.org/C18-1310.

139. Ece Takmaz, Sandro Pezzelle, Lisa Beinborn, and Raquel Fernández. Generating Image Descriptions via Sequential Cross-Modal Alignment Guided by Human Gaze. In *Proceedings of the 2020 Conference on Empirical Methods in Natural Language Processing (EMNLP)*, pages 4664–4677, Online, November 2020. Association for Computational Linguistics. https://doi.org/10.18653/v1/2020.emnlp-main.377. https://aclanthology.org/2020.emnlp-main.377.

140. Ekta Sood, Fabian Kögel, Florian Strohm, Prajit Dhar, and Andreas Bulling. VQA-MHUG: A gaze dataset to study multimodal neural attention in visual question answering. In *Proceedings of the 25th Conference on Computational Natural Language Learning*, pages 27–43, Online, November 2021. Association for Computational Linguistics. https://doi.org/10.18653/v1/2021.conll-1.3. https://aclanthology.org/2021.conll-1.3.

141. Sibo Dong, Justin Goldstein, and Grace Hui Yang. Gazby: Gaze-based bert model to incorpo-rate human attention in neural information retrieval. In *Proceedings of the 2022 ACM SIGIR International Conference on Theory of Information Retrieval*, ICTIR '22, page 182–192, New York, NY, USA, 2022. Association for Computing Machinery. ISBN 9781450394123. https://doi.org/10.1145/3539813.3545129. https://doi.org/10.1145/3539813.3545129.

142. Maria Barrett, Ana Valeria González-Garduño, Lea Frermann, and Anders Søgaard. Unsu-pervised induction of linguistic categories with records of reading, speaking, and writing. In *Proceedings of the 2018 Conference of the North American Chapter of the Association for Computational Linguistics: Human Language Technologies, Volume 1 (Long Papers)*, pages 2028–2038, New Orleans, Louisiana, June 2018b. Association for Computational Linguistics. https://doi.org/10.18653/v1/N18-1184. https://aclanthology.org/N18-1184.

143. Abhijit Mishra, Diptesh Kanojia, Seema Nagar, Kuntal Dey, and Pushpak Bhattacharyya. Leveraging cognitive features for sentiment analysis. In *Proceedings of the 20th SIGNLL Conference on Computational Natural Language Learning*, pages 156–166, Berlin, Germany, August 2016. Association for Computational Linguistics. https://doi.org/10.18653/v1/K16-1016. https://aclanthology.org/K16-1016.

144. Omid Rohanian, Shiva Taslimipoor, Victoria Yaneva, and Le An Ha. Using gaze data to pre-dict multiword expressions. In *Proceedings of the International Conference Recent Advances in Natural Language Processing, RANLP 2017*, pages 601–609, Varna, Bulgaria, Septem-ber 2017. INCOMA Ltd. https://doi.org/10.26615/978-954-452-049-6_078. https://doi.org/10.26615/978-954-452-049-6_078.

145. Xiaodi Zhang, Eric A. Maltbie, and Shella D. Keilholz. Spatiotemporal trajectories in resting-state fmri revealed by convolutional variational autoencoder. *NeuroImage*, 244: 118588, 2021a. ISSN 1053-8119. https://doi.org/10.1016/j.neuroimage.2021.118588. https://www.sciencedirect.com/science/article/pii/S1053811921008612.

146. Joachim Bingel, Maria Barrett, and Anders Søgaard. Extracting token-level signals of syntactic processing from fMRI–with an application to PoS induction. In *Proceedings of the 54th Annual Meeting of the Association for Computational Linguistics (Volume 1: Long Papers)*, pages 747–755, Berlin, Germany, August 2016. Association for Computational Linguistics. https://doi.org/10.18653/v1/P16-1071. https://aclanthology.org/P16-1071.

147. Nora Hollenstein, Cedric Renggli, Benjamin Glaus, Maria Barrett, Marius Troendle, Nicolas Langer, and Ce Zhang. Decoding eeg brain activity for multi-modal natural language processing. *Frontiers in Human Neuroscience*, page 378, 2021c.

148. Yuqi Ren and Deyi Xiong. CogAlign: Learning to align textual neural representations to cog-nitive language processing signals. In *Proceedings of the 59th Annual Meeting of the Associ-ation for Computational Linguistics and the 11th International Joint Conference on Natural Language Processing (Volume 1: Long Papers)*, pages 3758–3769, Online, August 2021. Asso-ciation for Computational Linguistics. https://doi.org/10.18653/v1/2021.acl-long.291. https://aclanthology.org/2021.acl-long.291.

149. Aarne Talman and Stergios Chatzikyriakidis. Testing the generalization power of neural net-work models across NLI benchmarks. In *Proceedings of the 2019 ACL Workshop Black-boxNLP: Analyzing and Interpreting Neural Networks for NLP*, pages 85–94, Florence, Italy, August 2019. Association for Computational Linguistics. https://doi.org/10.18653/v1/W19-4810. https://aclanthology.org/W19-4810.

150. Nora Hollenstein, Maria Barrett, and Lisa Beinborn. Towards best practices for leveraging human language processing signals for natural language processing. In *Proceedings of the Sec-ond Workshop on Linguistic and Neurocognitive Resources*, pages 15–27, Marseille, France, May 2020. European Language Resources Association. ISBN 979-10-95546-52-8. https://aclanthology.org/2020.lincr-1.3.

151. Nora Hollenstein. *Leveraging Cognitive Processing Signals for Natural Language Understanding*. PhD thesis, ETH Zurich, 2021.

152. Andrew A Krizhanovsky and Alexander V Smirnov. An approach to automated construction of a general-purpose lexical ontology based on wiktionary. *Journal of Computer and Systems Sciences International*, 52 (2): 215–225, 2013.

153. William Hart, Dolores Albarracín, Alice H Eagly, Inge Brechan, Matthew J Lindberg, and Lisa Merrill. Feeling validated versus being correct: a meta-analysis of selective exposure to information. *Psychological bulletin*, 135 (4): 555, 2009.

154. Kenneth I Forster. The potential for experimenter bias effects in word recognition experiments. *Memory & cognition*, 28 (7): 1109–1115, 2000.

155. Stan Szpakowicz. Last words: Failure is an orphan (let's adopt). *Computational Linguistics*, 36 (1), March 2010. https://doi.org/10.1162/coli.2010.36.1.36105. https://aclanthology.org/J10-1008.

156. Emiel van Miltenburg, Chris van der Lee, and Emiel Krahmer. Preregistering NLP research. In *Proceedings of the 2021 Conference of the North American Chapter of the Association for Computational Linguistics: Human Language Technologies*, pages 613–623, Online, June 2021. Association for Computational Linguistics. https://doi.org/10.18653/v1/2021.naacl-main.51. https://aclanthology.org/2021.naacl-main.51.

157. Zachary C Lipton and Jacob Steinhardt. Troubling trends in machine learning scholarship: Some ml papers suffer from flaws that could mislead the public and stymie future research. *Queue*, 17 (1): 45–77, 2019.

158. Lisa Beinborn and Rochelle Choenni. Semantic drift in multilingual representations. *Computational Linguistics*, 46 (3): 571–603, 2020. https://doi.org/10.1162/coli_a_00382. https://aclanthology.org/2020.cl-3.2.

159. Steffen Eger, Armin Hoenen, and Alexander Mehler. Language classification from bilingual word embedding graphs. In *Proceedings of COLING 2016, the 26th International Conference on Computational Linguistics: Technical Papers*, pages 3507–3518, Osaka, Japan, December 2016. The COLING 2016 Organizing Committee. https://aclanthology.org/C16-1331.

160. Bill Thompson, Sean Roberts, and Gary Lupyan. Quantifying semantic similarity across languages. In *Proceedings of the 40th Annual Conference of the Cognitive Science Society (CogSci 2018)*, 2018.

Procedural Strategies

6

The expressive power of human language is shaped by the fundamental principle of compositionality. We are able to combine the elementary units into larger blocks to derive novel expressions. While these combinations usually follow language-specific rules, the derivation of meaning from the parts is not straightforward. A simple example is the ambiguous interpretation of noun compounds: *olive oil* is oil made from olives but *baby oil* is not made from but for babies.

If we go beyond simple phrases, ambiguity increases, for example, due to underspecified semantic scope. In the sentence, *every baby oil contains an essential ingredient*, it remains unclear whether all oils contain the same essential ingredient. Another common example of procedural ambiguity is syntactic attachment. In the sentence *apply the oil on the shelf*, we need to integrate situational context to understand whether we should use the oil that can be found on the shelf or whether the shelf is the object onto which we should apply the oil.[1]

Humans are often able to directly resolve potential ambiguities by incorporating situational context to rule out unlikely alternatives. Implementing computational strategies for situated contextual disambiguation remains one of the big challenges towards cognitively more plausible models.

Procedural strategies in humans are characterized by an economic approach based on selective attention. During reading, function words are often skipped, while content words are fixated for a longer duration [1, 2]. The relative importance of tokens and phrases is weighted with respect to a comprehension objective. Reading processes for general understanding are different compared to skimming for information, checking grammar and style, or examining for factuality [3].

The procedural strategies of computational models are strongly determined by their architectural components. The various connecting mechanisms (such as pooling operations, tied

---

[1] In hierarchical syntactic approaches, the first interpretation is referred to as low attachment (to the noun phrase) and the second as high attachment (to the verb phrase).

© The Author(s), under exclusive license to Springer Nature Switzerland AG 2024   121
L. Beinborn and N. Hollenstein, *Cognitive Plausibility in Natural Language
Processing*, Synthesis Lectures on Human Language Technologies,
https://doi.org/10.1007/978-3-031-43260-6_6

parameters, residual connections, or attention heads) affect the information flow and influence how knowledge can be propagated and recombined through the layers. If the architecture is fixed, the choice of the input data and its representation restricts what the model can learn, and the target objective determines how information is weighted.

Recent analyses of language models indicate that they exhibit excellent strategies for memorization but struggle with generalization to unseen information [4]. Pimentel et al. [5] argue that these two aspects of language understanding do not need to be separated because humans make use of both skills. However, getting a better understanding of the procedural strategies that combine memorized knowledge with general abilities is important to anticipate the model's abilities on new tasks. Humans can apply their general language understanding to new tasks based on only a few instructive examples. Few-shot prompting simulates this scenario for computational language models [6]. Analyses that uncover the type of knowledge that is gained by the instructions can provide valuable information for a cognitively more plausible composition of the training data instead of simply increasing its size.

In this chapter, we first discuss analysis methods for understanding processes of relative importance and compositionality in the model using attribution signals. In the second part, we introduce methods to link these interpretability metrics to phenomena that we observe in cognitive signals. We then discuss how we can adjust the target objective and the inductive bias of the model towards cognitively more plausible procedural patterns.

6.1   Analyzing Computational Processing Signals

An important signal for analyzing processing strategies in neural networks are local attribution values. They can be extracted from neural networks to better understand the flow of information in the model.

Attribution signals are often characterized as a local signal that indicates the relative importance of input elements for the prediction of the output labels. Søgaard [7] explains that local attribution methods can also be applied globally. We can generalize from local attributions to processing patterns by calculating aggregate statistics over theory-driven linguistic categories of input tokens, for example, by grouping tokens by part-of-speech tag [8], by input position [9], or by gender [10]. This approach is comparable to the behavioral analysis of subpopulations (see Sect. 4.1.1.2). Ribeiro et al. [11] propose that the global properties of a model can best be understood by studying local explanations for a diverse, representative, non-redundant selection of instances.

Novel methods for interpretability of processing patterns are continuously emerging. Especially methods for quantifying the interaction between features can become relevant for analyzing cognitive plausibility in the future. For a more extensive discussion on the range of interpretability signals, see recent surveys by Søgaard [12] and Covert et al. [13].

We focus here on two signals that have recently been combined with cognitive data: attention values and gradient-based saliency.

6.1.1   Attention Values

The concept of attention plays an important role in contextualized language models. It was originally introduced as a weight matrix for a machine translation model that regulates which part of the input is taken into consideration when generating a translation [14]. In such encoder-decoder architectures, attention is motivated as a cognitively more plausible way to selectively attend to different parts of the input over time depending on which output token is being generated. The dynamic weighting mechanism is a simplification of human attention processes, but the general idea of modulating the information flow in neural networks can be intuitively mapped to selective cognitive attention during reading. Visualizations of attention can indicate the relative importance of input elements with respect to the target objective in a user-friendly way [15] and attention weights correlate well with cognitive patterns of visual saliency [16, 17].

6.1.1.1 Self-Attention

The architecture of transformer-based language models is usually limited to an encoder as they only operate on a single sequence (in contrast to classical sequence-to-sequence tasks such as machine translation or part-of-speech tagging). Within the decoder, self-attention is applied which is a more complex weighting mechanism that characterizes relationships between tokens (see Sect. 2.2.3). The self-attention mechanism has been motivated by engineering constraints because it facilitates parallelization and removes the left-to-right processing constraint. Consequently, the names of the key, query, and value matrices are borrowed from information retrieval systems and are not grounded in psychological concepts. Pretrained language models commonly apply not only a single attention head but multiple attention heads per layer to allow for more diverse attention perspectives [18]. It remains an open question how to best combine self-attention from different tokens, heads, and layers in a cognitively plausible way as multi-head attention can hardly be linked to a cognitive analog.

6.1.1.2 The Procedural Role of Attention

The procedural role of attention weights as an indicator of relative importance has been questioned [19–21]. For one, the calculated attention attends to input representations, not the input itself. As these representations are context-sensitive, they can mix in information from other inputs which makes it difficult to align attention information with input tokens [22]. Moreover, different attention distributions can lead to the same predictions [23], even when enforcing sparse attention distributions [24]. Although this is not necessarily implausible from a cognitive perspective (the prediction might simply be too obvious), it raises questions about the conclusions that can be drawn from attention patterns.

Recent analyses indicate that attention values might still produce usable explanations depending on the task and on how they are calculated [25]. Norm-based analyses of attention yield more plausible word alignments in machine translation [26]. Abnar and Zuidema [27] show that attention roll-out can capture the propagation of information from input tokens to intermediate hidden embeddings. They then interpret the attention graph as a flow network and compute maximum flow values from later attention layers to the input embedding layer. Unlike raw attentions weights, which consider token importance separately on each layer, this attention flow computes importance scores that account for mixing of information across layers and, thus, makes the identification of important tokens more transparent. According to Ethayarajh and Jurafsky [28], attention flow values can be interpreted as Shapley values.

6.1.2 Gradient-Based Saliency

Gradient-based saliency is determined to pinpoint those parts of the input that have the biggest influence on the prediction [29]. It is calculated as the gradient of the output corresponding to the correct prediction with respect to an input element.

If we have an input sequence $x_1, x_2, \ldots x_n$ with target class t, the gradient of the loss function with respect to an input token x_i can be calculated using the model output $f_t(x_1 : x_n)$. The gradient indicates the sensitivity of the model to the respective input token for predicting the target class. Gradient-based methods are considered to be a backward method [7] because the gradient is calculated from the last layer back to the input after the forward pass has been applied on the input—in contrast to the attention values that can be derived directly from the forward pass.

The terminology in the literature varies. Simonyan et al. [30] and Li et al. [31] use saliency to refer to the L2 norm of the gradient:

$$\text{saliency}(i, t) = \|\nabla_{\mathbf{x}_i} f_t(x_1 : x_n)\|_2 \tag{6.1}$$

Ancona et al. [32] and Bastings and Filippova [22] instead propose to refer to this value as sensitivity and use saliency as the dot product of the input and the gradient. More commonly, this second option is referred to as gradientXinput [33]. Conceptually, the gradient captures to what extent a change in the input leads to a change in the model prediction, and gradientXinput captures the contribution of the input to the model prediction. Saliency maps were first developed for image processing models to highlight the areas of the image that are discriminative with respect to the tested output class [30].

6.1.2.1 Saliency in Language Models

In language models, the objective function tries to predict a token based on its context. Accordingly, a saliency vector for a masked token indicates the importance of each of the tokens in the context for correctly predicting the masked token [34]. In order to determine the saliency of tokens with respect to the masked language modeling objective, we iteratively mask each token vector $\mathbf{x}_i$ in our input sequence $x_1, x_2, \ldots x_n$. Let $\mathbf{X}_i$ be the input matrix with $\mathbf{x}_i$ being masked. The model predicts a probability distribution over the vocabulary. The saliency of input token x_j is then calculated with respect to the prediction of the correct token $\mathbf{t}_i$.

$$s_{ij} = \|\nabla_{\mathbf{x}_j} f_{t_i}(\mathbf{X}_i)\|_2 \tag{6.2}$$

The relative importance of a token can then be determined by summing over its saliency for other tokens. As token-level saliency is affected by the length of the input sequence, the saliency values are commonly normalized by the sum or the maximum of all values in a sequence.

Gradient-based saliency does not work well when the gradient is zero (for example when using the ReLu activation function). Sundararajan et al. [35] calculate the relative sensitivity of the model to the input compared to a baseline to derive integrated gradients. In image classification, the baseline is usually a black image but it is still an open question how to best represent a neutral language input. Current approaches commonly resort to an all-zero vector or use the model-specific representation of the padding token or the unknown token.

6.1.2.2 Combining Gradients

Token-level saliency naively assumes that each token individually contributes to the prediction and ignores that language is compositional and that the relative importance of tokens can change depending on the context. Interpretability methods such as saliency are often exemplified for sentiment detection tasks. A negatively connotated adjective such as *disturbing* can be highly salient for predicting negative sentiment but it loses importance if it is already preceded by other negative adjectives. It might even be salient for positive sentiment if it is combined with other tokens (as in *The absence of disturbing neighbors made this campsite the perfect hideaway*).

Jacovi et al. [36] propose to isolate attribution effects using a contrastive interpretability approach that compares local attribution signals for minimal pairs of instances. Janizek et al. [37] calculate integrated Hessians over all pairs of input tokens to capture compositional relations such as negation. Their approach ignores any order constraints on compositional meaning which play an important role in many languages. Sikdar et al. [38] propose to calculate integrated directional gradients over groups of features. In practice, this means that they apply a constituency parser on the model input and simply sum the integrated gradient values within a phrase. The resulting saliency trees facilitate interpretation of the model behavior but the approach is constrained by the availability and the performance of the parser.

The non-linear combination of input elements is one of the main driving forces behind the success of neural models [39]. Better understanding the patterns underlying these non-linear processes remains a big challenge for methodological interpretability research but will be a key factor in developing cognitively more plausible models.

6.2   Testing Processing Strategies

Testing the cognitive plausibility of neural language processing strategies can follow two lines of experimental evaluation: correlational and predictive. In a correlational setup, we use cognitive data to analyze the similarity between language model responses and human responses. In a predictive setup, we use language models to predict human responses. Both directions can be analyzed quantitatively and are strongly related. They only differ in the understanding of the role of the computational model.

The goal is to learn more about the procedural strategies of the computational model with the help of cognitive data. In the correlational setup, we interpret the cognitive data as the ground truth response and analyze the cognitive plausibility of different modeling setups. This perspective is more common in natural language processing, where we assume that closer similarity between cognitive signals and computational representations are reflective of a model's processing strategies. In the predictive setup, the computational model explicitly operationalizes theoretical assumptions about processing strategies. We assume that human language processing signals serve as a gold standard for prediction. If language models can accurately predict cognitive signals, they are assumed to rely on comparable language processing strategies as humans. This perspective is more common in psycholinguistics. The goal is to better explain human processing strategies with the help of computational models.

We believe that both directions are fruitful and need to be combined. Laverghetta Jr. et al. [40] propose that language models can be used for predicting the outcome of psychometric tests and facilitate the selection of meaningful test items. However, we need to be careful not to get trapped in circular argumentation. We cannot claim that a model is more cognitively plausible if it exhibits similar processing patterns than humans while explaining the human patterns with the help of exactly the same model.

6.2.1   Relative Importance

When children learn to read, they first focus on each word individually and gradually learn to anticipate frequent patterns [41]. Determining the relative importance of the elements in a sentence is a key factor for effortless natural language understanding. More experienced readers are able to completely skip words that can be inferred from the context and focus on the more relevant words of a sentence [42]. We optimize a tradeoff between the precision of comprehension and the economy of attention [43]. In other words, we want to represent the meaning conveyed by the input as accurately as possible while limiting our cognitive effort.

The sentence processing effort can be approximated indirectly using a range of metrics such as response times in reading comprehension experiments [44], word-by-word processing duration in self-paced reading [45], voltage changes in electroencephalography recordings [46], or eye movement recordings during reading [47].

6.2.1.1 Token-Level Linking of Gaze Patterns

Gaze patterns captured by eye-tracking technology indicate how a reader processes a text. It has been shown that fixation duration on a token and the relative importance of that token for comprehension are strongly correlated in natural reading [48]. It is thus possible to establish a token-level linking of importance and compare it to attribution signals from computational language models.

For example, the gradient-based saliency values of a BERT model show that the token *ingredients* is relatively more important than the other tokens in the sentence *For the most part the ingredients are there*. The mean fixation duration values for reading the same sentence indicate a similar pattern (see Fig. 6.1).

Such parallels between human and computational processing patterns have recently been found in multiple analyses. Sood et al. [49] measure the correlation between attention in neural language models finetuned for a document-level question-answering task and find that last-layer attention in a transformer model deviates strongly from human fixation patterns in contrast to attention in a convolutional and an LSTM model. In our own work, we confirm the

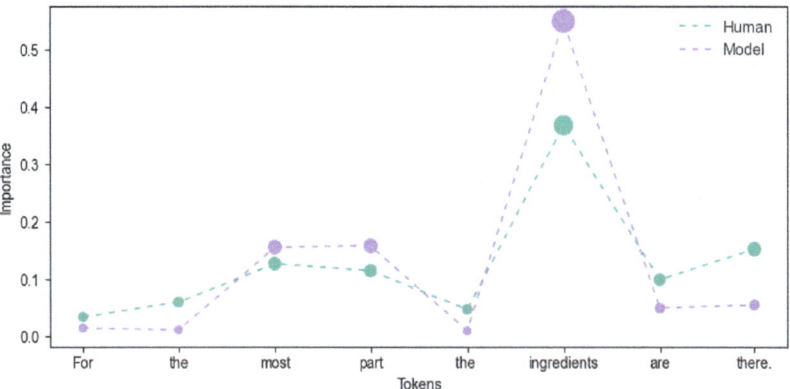

Fig. 6.1 Example of gradient-based saliency values of a BERT model compared to human relative fixation duration captured with eye-tracking. The example was retrieved from the code by Hollenstein and Beinborn [8]

weak correlation for last-layer attention in transformer-based language models and propose to investigate gradient-based saliency instead [8]. We find that it correlates strongly with relative fixation duration in natural reading comprehension and identify comparable effects of frequency and word class in human and computational processing.

Recent analyses of attention weights indicate that the last layer is not a good choice for token-level effects as these seem to be more strongly represented in lower layers [50]. Bensemann et al. [51] compare a range of attention configurations by correlating them with human signals from multiple eye-tracking corpora. They find that the correlation is generally stronger in the lower layers of the model and that the first layer is usually closest to human patterns. They observe that model attention corresponds more strongly to human attention during reading for general comprehension than for more task-specific reading.

Eberle et al. [52] go deeper into task-specific reading and analyze relative importance of tokens for sentiment analysis and relation extraction. They find that attention flow over layers correlates well with human gaze patterns but that the neural model processes proper nouns and unexpected words differently than humans.

In these studies, a higher correlation between attribution patterns and cognitive signals is usually interpreted as higher cognitive plausibility of the model. However, the influence of the choice of the attribution methods, the underlying model architecture, and the type of reading task on the observed cognitive plausibility of the model remains an open research question. The conclusions drawn from these approaches are closely linked to the distinction between faithfulness and plausibility of interpretability methods [53]. An attribution method might correlate well with human processing patterns but if it is not faithful to the computational processes in the model, it cannot inform about the model's strengths and weaknesses.

6.2.1.2 Cross-Lingual and Individual Differences

It is a challenging task to generalize importance patterns across languages and across individuals. Gaze patterns are influenced by surface characteristics such as orthography and word length [54]. For example, reading text in Chinese characters results in longer fixations and narrower saccades because the script is visually denser than texts using the Latin alphabet [55]. It currently remains an open research question whether such superficial differences converge to more stable reading patterns on a deeper processing level. Semantically aligned texts are read at a similar speed in Chinese, English, and Finnish [55] and similar predictability effects have been reported in multiple languages [56, 57].

Pouw et al. [58] show that sentence-level patterns of cognitive processing complexity can be projected across 13 typologically diverse languages. Their control experiment with randomized word order indicates that a model finetuned on gaze data develops an increased sensitivity to structural information. Morger et al. [59] conduct cross-lingual analyses of relative importance on the token level and find that the observed correlation between human importance and attribution values generally remains stable for four languages (English, Dutch, German, and Russian). They observe a slightly better fit for monolingual models over multilingual models and confirm that last-layer attention does not match human patterns. The differences between gradient-based saliency, first-layer attention, and attention flow are relatively small in their experiments.

Most analyses average eye-tracking metrics over readers to obtain a more robust indication of human reading behavior. This approach disregards the fact that reading is a highly individual process that is dependent on cognitive factors and experience (see Sect. 3.4). Brandl and Hollenstein [60] investigate differences between individual readers. They show that the correlation between attention values in multilingual BERT and gaze patterns in the MECO corpus vary strongly between languages (Spearman correlation coefficients ranging between 0.27 for English to 0.53 for Greek). Interestingly, a higher correlation can be found for participants who perform in-depth reading (which is characterized by a lower skipping rate and longer fixation durations) in contrast to more shallow processing during fast reading. This observation can be linked to the finding that computational processing patterns correspond better to general language comprehension than to task-specific reading.

6.2.1.3 Predicting Gaze Patterns

Instead of exploring human importance patterns in computational models, we can test the predictive power of our model to project human reading behavior for unseen stimuli [61]. The Cognitive Modeling and Computational Linguistics (CMCL) Workshop organized a shared task that challenged researchers to train a model to predict word-level eye-tracking metrics both in monolingual [62] and multilingual scenarios [63]. The goal of both tasks was to predict gaze features such as first and total fixation duration based on training data from eye-tracking datasets recorded during sentence comprehension.

For multilingual prediction, most approaches finetune transformer-based multilingual models [64] or use adapters to transfer information between languages [65]. Interestingly, enriching neural models with explicit linguistic information about the length and frequency of tokens improves the prediction performance [59, 66]. This could indicate that this information is not sufficiently encoded in the model representations or that it needs to be more directly integrated as inductive bias so that the model learns to focus more on this information.

Most prediction experiments are conducted on the sentence level while the eye-tracking signals are obtained from participants reading longer texts. Wiechmann et al. [67] show that explicitly adding information from the preceding sentence helps to account for spill-over effects across sentence boundaries.

Relative importance of a token for sentence processing is an underspecified term that encompasses several related local processing effects such as relevance for understanding the sentence, difficulty, and novelty of a token within the context, semantic and syntactic surprisal, or domain-specificity of a token.

6.2.2 Local Processing Effects

When we process language, we incrementally integrate new information into our representation. At the same time, we build up expectations about upcoming information. If our expectations turn out to be wrong, we need to readjust our representation to accommodate the unexpected piece of information. Recall, the example *The old train the young* in Sect. 2.1.2. When we encounter the second *the*, we need to repair the dominant interpretation of *train* being a noun.[2] This phenomenon is called surprisal and it can be linked to processing difficulty [69, 70]. Futrell and Levy [71] summarize that according to the surprisal theory, "the processing cost of a word is asserted to be proportional to the extent to which one must change one's beliefs given that word". They approximate surprisal with a probabilistic language model and show that it is proportional to reading times in self-paced reading.

Linzen and Jaeger [45] compare two main hypotheses of local processing difficulty: the competition hypothesis assumes that sentence processing difficulty is increased if we need to maintain a large number of potential interpretations with comparable probability. For example, a sentence beginning with *We admit* can be continued by a noun phrase (*our mistake*) or by a sentential complement (*that we were wrong*). The entropy reduction hypothesis instead assumes that the processing load is increased by a reduction in uncertainty. According to this hypothesis, it is not the maintenance of multiple prediction candidates caused by reading *admit* that causes processing complexity but rather the elimination of candidates caused by reading the following word (*our* vs *that*). They analyze self-paced reading times and find that both surprisal (encountering an unexpected word) and entropy reduction (encountering a disambiguating word) account for the phenomena in the data but that increased processing difficulty due to competition (encountering a word that increases uncertainty) cannot be corroborated by the measured reading times.

More recent analyses indicate that surprisal alone underestimates

6.2.2.1 Forgetting

It has been shown that processing difficulty increases with the size of the context window that needs to be kept active for understanding [72]. Sentences are harder to process if they contain long-distance dependencies, for example, due to embedded relative clauses [73]. Interestingly, there seems to be a limit on the processing window for open dependencies. The phenomenon of structural forgetting describes the observation by Gibson and Thomas [74] that humans find the shorter ungrammatical version of a sentence (example 1) at least as acceptable as the more complex grammatical construction that requires long-distance agreement (example 2). This phenomenon is also known as the grammaticality illusion [75]

---

[2] The dominance of the noun reading is due to the English processing preference of late closure [68]. We prefer to keep the current phrase open instead of closing the noun phrase after *old* and opening a verb phrase.

1. *The apartment that the maid who the service had sent over was well decorated.*
2. *The apartment that the maid who the service had sent over was cleaning every week was well decorated.*

Futrell et al. [76] assume that humans maintain an incremental syntactic state during sentence processing. They use psycholinguistic experimental techniques developed for studying humans to test syntactic generalization in computational models and find evidence for a basic incremental state of the syntactic representation. Their analyses indicate that models can only develop a sensitivity to subtle lexical cues that signal changes in the syntactic state if they are trained on large datasets containing a sufficient range of such phenomena.

The sentence processing experiments described above used English stimuli but increasing evidence is found that structural preferences are language-specific. Frank et al. [77] show that structural forgetting observed for double-embedded relative clauses in English sentences cannot be confirmed for self-paced reading of Dutch and German sentences. However, the Dutch and German readers show similar behavior of structural forgetting when reading the English stimuli indicating that the effect is language-specific. They claim that cross-lingual differences in the distribution of the underlying statistical patterns affect the processing load. When they train a computational model on English and Dutch data simultaneously, it accurately predicts the cross-lingual differences in human processing.

Gildea and Jaeger [78] propose that languages evolve over time to increase processing efficiency. They find that word order constraints are optimized to minimize trigram surprisal of words. Gibson et al. [79] refer to this phenomenon as communicative efficiency and provide a survey of information-theoretic evidence for lexical and syntactic changes toward reducing cognitive load. Such information-theoretic assumptions are harder to map to neural language models than to purely statistical approaches. Another problem is that the training data for language models has become so large that it becomes infeasible to calculate frequency statistics over the full dataset. However, the same is true for human participants in psycholinguistic studies: we cannot quantify the properties of the linguistic input they have been exposed to in the past.

6.2.2.2 Predicting M/EEG Responses

In EEG signals, surprisal is often reflected by an N400 component (see Sect. 3.1). The strength of this component is shaped by the syntactic and semantic properties of the context. Frank and Willems [80] distinguish between syntagmatic relations to refer to words that are linked sequentially (i.e., they often occur next to each other) and paradigmatic relations to refer to words that are semantically related (i.e., they occur in similar contexts). They calculate surprisal and semantic distance in sentences to predict changes in electrode potential in EEG signals of English and Dutch readers. They find that both metrics can be linked to highly similar observations in the N400 response indicating cognitive interdependencies between the two phenomena.

Research on the prediction of EEG signals is often driven by determining which computational mechanisms better anticipate human responses. While this debate focused on comparing symbolic and probabilistic approaches in the past, the current focus has been narrowed down to comparing recurrent language models to transformer-based language models. The main difference is that recurrent models process the input incrementally and can be more easily mapped to memory-related phenomena influenced by processing distance such as forgetting and information locality. Transformer models, in contrast, can access the entire context window (i.e., 1,024 tokens in GPT-2 and 512 tokens in BERT). Jat et al. [81] find that the transformer-based model BERT better predicts MEG signals than recurrent neural networks. Merkx and Frank [82] and Michaelov et al. [83] come to similar conclusions when predicting EEG signals. Their analyses suggest that semantic aspects of human language comprehension may be better modeled with transformer-based surprisal whereas aspects associated with limited working memory may be better approximated by surprisal in recurrent LMs. Similar tendencies have been observed for predicting reading times [84, 85] and for fMRI and eCOG data [86]. It should be noted that we provide merely a high-level summary of the findings on predicting brain activity data. We suggest cautiously examining the individual results as the experiments are run on different datasets, using differing and not always fully transparent experimental choices and varied evaluation metrics. As the observed effects are usually modest, more research on publicly available datasets and standardized evaluation procedures are required before we can reliably generalize these findings. Additionally, the training data of the language models under study varies with respect to size and quality. This imparity might influence the effects that have been attributed to the model architecture by the authors.

Another important debate focuses on the question of whether a hierarchical representation is required for language processing [87]. Traditional grammar formalisms use tree structures to distinguish between different syntactic representations of a sentence, for example between high and low attachment of a prepositional phrase (recall the oil example at the beginning of the chapter). Most neural language models instead interpret sentences as a sequence but still perform well on syntactic generalization. Frank et al. [46] find that flat and data-driven language models better fit EEG responses than a hierarchical phrase structure grammar. In their opinion, these findings suggest "that readers do not make use of a sentence's hierarchical structure for generating expectations about the upcoming word." Linzen and Baroni [88] provide a detailed overview of the discussion around hierarchical processing. They summarize several studies including Hale et al. [89], McCoy et al. [90], and Futrell et al. [76] by stating:

[W]hen DNNs are explicitly constrained to process words in an order dictated by a parse tree, rather than from left to right, their syntactic behavior more closely matches that of humans: they require less data than standard DNNs to acquire certain generalizations, and they generalize in a human-like way to new syntactic structures.

We think that it is too early to conclude that one approach is clearly better than the other as only a tiny section of the experimental hypothesis space has been explored to date. More interdisciplinary research is required to control for all the potential confounding factors related to the cognitive data sample, the language and the stimuli under study, the training data and the architectural details of the model, the evaluation setup, and the confirmation bias of the experimenter.

6.2.2.3 Priming

Psycholinguistic studies show a robust effect of priming: lexical access to a concept is facilitated, when the context contains a semantically related or associated concept [91]. For example, producing *pineapple* is facilitated if the context contains tokens that refer directly to other tropical fruit (*mango* or *coconut*) but also if it contains tokens that we associate more freely with tropical fruit such as *cocktail* or *summer*.[3] Misra et al. [92] use English lexical stimuli that elicit priming effects in humans to query the contextualized language model BERT. They find that the model predicts a word with greater probability when the context includes a related word versus an unrelated one. For example, predicting the word *pilot* in the sentence *I want to become a ____* is more likely when the preceding context contains *airplane* rather than *table*. Their findings indicate that priming can be induced by a large range of relationships between tokens and that synonymy, antonymy, and category relations cause the most pronounced differences.

Priming can also occur on the syntactic level. Humans prefer certain syntactic constructions when they have been exposed to them more frequently [93]. Prasad et al. [94] compare priming effects for different types of relative clauses by adapting a recurrent language model. In their methodology, adapting refers to continual pretraining of the model using only sentences with a fixed syntactic structure.

1. Reduced: *The pizza my friend ordered was delicious.*
2. Full: *The pizza that my friend ordered was delicious.*

They assume that a model that has been adapted to reduced relative clauses as in example 1 will assign lower surprisal to the reduced form than an unadapted model or a model adapted to the full form in example 2. They confirm the syntactic priming effect for all seven types of relative clauses in their study.

The results of the priming studies indicate that the inductive bias of the language models can easily be adapted. This is conceptually comparable to humans who adapt to the conditions of a task or to a target audience. We use language differently when communicating with children than when engaging in a professional dispute with peers. Once we know how a model diverges from human strategies for solving a specific task, we can nudge the model

---

[3] Some people might also have a weird priming association between pizza and pineapple, but that's another story.

towards cognitively more plausible processing. Pilot studies indicate that language models which are adapted to better align with cognitive signals might even yield quantitative benefits on downstream tasks. Fyshe et al. [95] find that word representations trained on brain activity are more generalizable to unseen words and Schwartz et al. [96] propose that tuning language models on MEG data can be beneficial for the performance on a subset of the GLUE tasks [97].

Nevertheless, we need to be aware that priming can also have negative effects. It is well known, that human answers to a survey are influenced by the form, the order, and the context of the questions [98]. Kassner and Schütze [99] show that language models can easily be tricked intro wrong answers. The language model BERT-large incorrectly fills the gap in the prompt "*Platonism is named after [MASK]*" if the distracting name *Cicero* precedes the sentence. This effect is called mis-priming and reveals an important risk. When applying a language model in practice, it might adapt to the unexpected and sometimes even malicious behavior of its users [100]. To take effective counter measures, more research on uncovering the procedural biases of language models is required.

6.3   Towards Cognitively Plausible Processing

When humans have acquired fluent reading skills, we are able to use them to solve all kinds of problems with minimal instruction. We are able to integrate the syntactic and semantic information in a text and can summarize it, extract facts, interpret its tone or stylistic choices, and connect it with our background knowledge. Recent evaluation benchmarks evaluate models on a variety of tasks to assess their generalizability (see Sect. 4.2.1). The tasks cover diverse aspects of language understanding but each task is represented by a separate dataset with its own sample distribution. During finetuning, the model learns the procedural patterns that are relevant for solving the task for this particular dataset. This means that a good performance might be obtained by learning latent biases in the dataset that facilitate shortcuts (see Sect. 4.1.2). Closing the gap within one dataset with a matching contrast set can be challenging [101]. As an alternative, we discuss how the inductive bias of a model can be modified towards cognitively more plausible models.

We first discuss two learning strategies that aim at exploiting the complementary characteristics of datasets designed for tasks that are conceptually related: multi-task learning and transfer learning. Multi-task learning refers to a change in the architecture and the target objective so that multiple tasks are learned in parallel. Transfer learning refers to a sequential setup: the model first learns general patterns for a related task (usually one for which more data is available) and then ideally only needs to learn subtle adjustments to represent the target task. Both strategies are based on the idea that a model that is able to solve multiple tasks is less likely to overfit to superficial characteristics of the training data. We then discuss approaches that examine the target objective of a language model from a more theoretical linguistic perspective.

6.3.1   Multitask Learning

Multitask learning is an established technique for increasing the robustness of a model by optimizing for more than one task using shared representations [102]. In practice, a multitude of architectural choices and loss combinations can work well for multi-task learning [103]. In natural language processing, we often distinguish between a main task and an auxiliary task when applying multitask learning. The intuition behind that setup is that we can improve the model's performance on the main task by explicitly optimizing for an auxiliary objective. We thus nudge the model towards picking up patterns that we consider to be useful for the main task but that are more explicit in the data for the auxiliary task. For example, being able to predict the part-of-speech class of a word might be useful for predicting named entities because they frequently occur as noun compounds [104]. As the model parameters are shared across tasks, it is important that the two tasks are meaningfully related.

For a cognitively more plausible model, we want to integrate information on human problem-solving strategies. We can do this explicitly by making assumptions about human behavior (see Chap. 4) or we can provide more implicit information encoded in cognitive signals. If a model is able to predict cognitive data of human language processing, it might learn cognitively more plausible patterns for extracting information from text [105]. For example, González-Garduño and Søgaard [106] show that a model can better predict the readability of a sentence if it simultaneously optimizes the auxiliary task of predicting eye tracking patterns of human readers. Eye tracking information also helps in learning patterns for sentence compression [107] and part-of-speech tagging [108].

Cognitive signals provide information about the selective attention processes that humans apply. In Sect. 5.3.2.1, we describe how eye-tracking patterns can be used as a proxy for visual saliency and in Sect. 6.2.1.1, we introduce cognitively inspired analyses of relative importance in computational text processing. When we simply fuse the cognitive signal to the input (see Sect. 5.3.2.1), we leave it up to the model how to integrate the information. Instead, we could use it more directly by regulating the attention weights which are meant to steer selective attention in models. McGuire and Tomuro [109] improve a model for relation extraction by using a joint loss that monitors the attention weights assigned to the input tokens and compare them to an attention score from eye tracking and EEG data. When using cognitive signals during training a model, we face two main problems: (i) the available datasets are too small for the large parameter space of neural models (ii) cognitive data is not available during test time. To account for these challenges, Barrett et al. [110] propose a training regime that alternates between the main NLP task and the auxiliary task of learning attention weights using eye tracking data. The auxiliary batches are used to update the attention weights and the task-specific batches optimize all other parameters. Their approach makes it possible to use disjoint datasets (i.e., the reading stimuli for the eye tracking signal do not need to be parallel to the task data) and still train the model on two related objectives in parallel. Due to the alternating procedure, the model does not require cognitive signals for inference. Muttenthaler et al. [111] show that this approach can also be used with EEG data.

6.3.2 Transfer Learning

Transfer learning focuses on applying knowledge gained while solving one problem to a different but related problem. For example, a model trained to do sentiment analysis can subsequently be finetuned to a sarcasm detection task. Transfer learning techniques include any method that re-purposes a model for a different task. Rather than learning tasks simultaneously as in multi-task learning, transfer learning is conducted sequentially.

In a typical transfer learning scenario with cognitive data, we first train a model to predict cognitive features and then re-use the frozen hidden layers in a new network to output predictions for a natural language processing task. For example, Evaldo Leal et al. [112] propose to first predict eye tracking features and then transfer the frozen hidden layers to a network that predicts text readability for Brazilian Portuguese texts.

Any finetuning of a pretrained language model can be seen as transfer learning because the model is first optimized for a language modeling objective (e.g. masked word prediction) and then trained on task-specific data. Schwartz et al. [96] introduce an additional step and first fine-tune a language model to predict MEG data and in a second step, adapt it to the tasks in the GLUE evaluation benchmark [97]. They show that this adaptation does not diminish the performance. More detailed instance-level evaluation is required to analyze if the cognitive plausibility of the model predictions increases.

Toneva and Wehbe [113] serendipitously find that changing the attention in lower layers of BERT to a uniform pattern leads not only to improved prediction performance of fMRI signals but surprisingly also yields an increased performance on task that require syntactic understanding. This is an example of how insights derived from experiments with cognitive signals can be transferred to develop better NLP models and to question established methodological choices.

To fruitfully combine NLP tasks for cognitively plausible transfer learning, it is important to understand how tasks are related to each other and to what extent they rely on shared processing patterns. Luo et al. [114] aim at finding relations between tasks using fMRI signals to determine promising pairs for transfer. Conceptually, transfer learning is closely related to curriculum learning. Ideally, a model first learns general patterns on a cognitively easier task before attempting a more challenging task.

A challenge for transfer learning is the phenomenon of catastrophic forgetting [115]. Catastrophic forgetting refers to the phenomenon that a neural model which is trained on two tasks sequentially forgets information relevant to the first task while optimizing its representations for the second task. Catastrophic forgetting is a crucial problem for both transfer learning and curriculum learning as proficiency in the conceptually easier tasks should not be lost when learning to solve the harder tasks (see Sect. 4.3.2). We expect that finding a better weight adaptation regime for sequential learning can be facilitated by a better understanding of how humans assign relative importance to new information. Following an economic processing approach requires taking some shortcuts to build more robust representations without losing the core information.

6.3.3   Integrating Linguistic Information

Pre-trained language models learn a surprising amount of linguistic knowledge, especially syntactic knowledge, without any explicit linguistic supervision [116, 117] but they require large amounts of data to generalize to complex linguistic phenomena [118]. van Schijndel et al. [119] show that simply increasing the training data does not help for all cases of syntactic generalization. Furthermore, analyses by Carlini et al. [120] indicate that larger portions of training data reinforce undesirable memorization. In order to arrive at more data-efficient models, we need to find ways to integrate linguistic constraints into the training procedure. Such hybrid models that combine neural and symbolic approaches are a promising research direction toward cognitively plausible generalization and causal inference. For example, Dyer et al. [121] combine a language model with a parser to explicitly model nested, hierarchical relationships among words and phrases. Abdou et al. [122] enrich the masked language modeling objective in BERT with an additional structural attention constraint using annotations of syntactic dependency relations and minimal recursion semantics. They show that the linguistic information improves the performance of a linear decoder that maps the representations of stimuli sentences to their brain recordings but report discrepancies between word-level and sentence-level results.

The uniform information density (UID) hypothesis is a popular psycholinguistic theory for explaining empirically observed syntactic, morphological, and prosodic choices. It posits that speakers tend to distribute information uniformly across a linguistic signal. For example, many speakers will prefer the utterance *My boss confirmed that we are crazy* over *My boss confirmed we are crazy* because *we* exhibits disproportionately high information density if the relativizer *that* is missing [123]. Including *that* leads to a more uniform distribution of the per-word information density. The information conveyed by a word can be approximated by the surprisal associated with it (see Chap. 2).

Wei et al. [124] augment the training objective of a transformer-based language model with a regularizer that enforces the UID hypothesis. They find that models with UID regularization obtain better perplexity when only limited data is available and generate lexically more diverse sentences. Their results are robust across a range of languages (including Swahili, Tagalog, Finnish, Czech, and English) and suggest that UID can be a cognitively more plausible inductive bias for language modeling. Interestingly, this finding does not seem to generalize to all languages as the UID observation seems to be related to word order preferences [125].

Kuribayashi et al. [126] align surprisal metrics of language models with eye tracking data of Japanese reading. They find that lower language model perplexity does not indicate better performance for modeling reading behavior in contrast to their findings for English. Further analysis shows that the Japanese gaze patterns indeed deviate from the UID hypothesis and that language models that perform better on predicting the reading data are more sensitive to the syntactic category of tokens and distribute information less uniformly. They conclude that lower perplexity of a model does not necessarily yield a cognitively more plausible model. Oh and Schuler [127] similarly find that large transformer-based pretrained language models with more parameters and lower perplexity yield surprisal estimates that are less predictive of human reading times. They assume that this observation is due to the tendency of these models to memorize sequences during training. Their findings are in line with earlier analyses showing that surprisal systematically underpredicts human behavioral responses [128–130]. These findings indicate that the language modeling objective might be insufficient to capture a generalizable understanding of syntactic complexity. While low perplexity on an evaluation set undoubtedly reflects some level of fit to the training data, it does not give us a finegrained view of the linguistic attributes that the model has learned [131].

In order to get to such a more finegrained understanding of processing patterns, Davis and van Schijndel [132] investigate how pronoun resolution preferences differ for English, Chinese, Italian, and Spanish. They investigate implicit causality verbs that trigger either a subject bias or an object bias and find that the computational models acquire the same preferences as humans for English and Chinese, but not for Italian and Spanish. Their analyses indicate that the model failure can be explained by its inability to learn the correct interaction between competing constraints and that the training procedure can be adapted to better capture the phenomenon. These findings strengthen the argument that a one-size-fits-all solution for the training parameters and the architectural choices of models cannot account for cross-lingual variability.

6.4   Ethical Aspects

We have seen that the procedural strategies of contextualized language models can be adapted using finetuning and priming techniques. Such adaptations can be exploited for harmful scenarios. Experiments with chatbots have shown that language models can be subject to emergent bias which occurs as a result of use and interaction with real users [133]. If users teach a chatbot to be racist [134], the responsibility for the adaptation cannot be clearly attributed anymore as it occurs in a distributed and not necessarily intentional fashion.

Such ethical questions about accountability and fairness cannot be addressed by natural language processing researchers alone as they are embedded in more general sociological, legal, and psychological debates [135]. As academics, we are responsible for integrating ethical considerations in educational curricula [136] to increase the sensitivity of future users and developers of cognitively plausible models to ethical issues [137]. This is not an easy task. Hagendorff [138] criticizes that the dominant topics in guidelines about ethical AI are the ones that can be addressed technologically and that the wider societal and economic context in which models are applied is currently disregarded. Mittelstadt [139] notes that we are still far from consensus about high-level ethical principles in the application of computational models due to strong political and normative disagreement. Hickok [140] adds that we need to increase the diversity of perspectives because the interpretation of ethical values is culture-sensitive. While there is general agreement in the research community that computational models should not do harm, it remains an open question how to translate this into actionable guidelines.

Many ethical guidelines highlight the importance of transparency of AI models which is closely connected to principles of communication and traceability. Clearly communicating to users that they are interacting with an AI system can be easily implemented and is of utmost importance when developing cognitively more plausible models. The obligations that come with the principle of traceability (often more strongly put as explainability) are harder to operationalize. While the UN principles clearly demand that users have "access to the reasons for a decision and the logic involved" [141], the EU guidelines acknowledge that "an explanation as to why a model has generated a particular output or decision (and what combination of input factors contributed to that) is not always possible" [142]. As a workaround they demand traceability of the model and human agency with respect to the decisions. Traceability can only be achieved with a high level of transparency in the documentation of data gathering, data labeling, and algorithmic implementation. Human agency requires that users are "given the knowledge and tools to comprehend and interact with AI systems to a satisfactory degree and, where possible, be enabled to reasonably self-assess or challenge the system". It is our responsibility as researchers to contribute to these aspects by clearly documenting our experimental choices and by unveiling the mechanisms underlying the procedural strategies of our models.

References

1. Keith Rayner, Sara C Sereno, Robin K Morris, A Rene Schmauder, and Charles Clifton Jr. Eye movements and on-line language comprehension processes. *Language and Cognitive Processes*, 4(3-4):SI21–SI49, 1989.
2. A René Schmauder, Robin K Morris, and David V Poynor. Lexical processing and text integration of function and content words: Evidence from priming and eye fixations. *Memory & Cognition*, 28(7):1098–1108, 2000.
3. Ralf Biedert, Jörn Hees, Andreas Dengel, and Georg Buscher. A robust realtime reading-skimming classifier. In *Proceedings of the symposium on eye tracking research and applications*, pages 123–130, 2012.
4. Aparna Elangovan, Jiayuan He, and Karin Verspoor. Memorization vs. generalization : Quantifying data leakage in NLP performance evaluation. In *Proceedings of the 16th Conference of the European Chapter of the Association for Computational Linguistics: Main Volume*, pages

1325–1335, Online, April 2021. Association for Computational Linguistics. https://doi.org/10. 18653/v1/2021.eacl-main.113. https://aclanthology.org/2021.eacl-main.113.

5. Tiago Pimentel, Josef Valvoda, Rowan Hall Maudslay, Ran Zmigrod, Adina Williams, and Ryan Cotterell. Information-theoretic probing for linguistic structure. In *Proceedings of the 58th Annual Meeting of the Association for Computational Linguistics*, pages 4609–4622, Online, July 2020. Association for Computational Linguistics. https://doi.org/10.18653/v1/2020.acl-main.420. https://aclanthology.org/2020.acl-main.420.

6. Pengfei Liu, Weizhe Yuan, Jinlan Fu, Zhengbao Jiang, Hiroaki Hayashi, and Graham Neubig. Pre-train, prompt, and predict: A systematic survey of prompting methods in natural language processing. *ACM Comput. Surv.*, 55(9), jan 2023. ISSN 0360-0300. https://doi.org/10.1145/3560815. https://doi.org/10.1145/3560815.

7. Anders Søgaard. Explainable natural language processing. *Synthesis Lectures on Human Language Technologies*, 14(3):1–123, 2021.

8. Nora Hollenstein and Lisa Beinborn. Relative importance in sentence processing. In *Proceedings of the 59th Annual Meeting of the Association for Computational Linguistics and the 11th International Joint Conference on Natural Language Processing (Volume 2: Short Papers)*, pages 141–150, Online, August 2021. Association for Computational Linguistics. https://doi.org/10.18653/v1/2021.acl-short.19. https://aclanthology.org/2021.acl-short.19.

9. Sumit Chopra, Michael Auli, and Alexander M. Rush. Abstractive sentence summarization with attentive recurrent neural networks. In *Proceedings of the 2016 Conference of the North American Chapter of the Association for Computational Linguistics: Human Language Technologies*, pages 93–98, San Diego, California, June 2016. Association for Computational Linguistics. https://doi.org/10.18653/v1/N16-1012. https://aclanthology.org/N16-1012.

10. Ji Ho Park, Jamin Shin, and Pascale Fung. Reducing gender bias in abusive language detection. In *Proceedings of the 2018 Conference on Empirical Methods in Natural Language Processing*, pages 2799–2804, Brussels, Belgium, October-November 2018. Association for Computational Linguistics. https://doi.org/10.18653/v1/D18-1302. https://aclanthology.org/D18-1302.

11. Marco Ribeiro, Sameer Singh, and Carlos Guestrin. "why should I trust you?": Explaining the predictions of any classifier. In *Proceedings of the 2016 Conference of the North American Chapter of the Association for Computational Linguistics: Demonstrations*, pages 97–101, San Diego, California, June 2016. Association for Computational Linguistics. https://doi.org/10. 18653/v1/N16-3020. https://aclanthology.org/N16-3020.

12. Anders Søgaard. Neural speed reading audited. In *Findings of the Association for Computational Linguistics: EMNLP 2020*, pages 148–153, Online, November 2020. Association for Computational Linguistics. https://doi.org/10.18653/v1/2020.findings-emnlp.14. https://aclanthology.org/2020.findings-emnlp.14.

13. Ian Covert, Scott M Lundberg, and Su-In Lee. Explaining by removing: A unified framework for model explanation. *J. Mach. Learn. Res.*, 22:209–1, 2021.

14. Dzmitry Bahdanau, Kyung Hyun Cho, and Yoshua Bengio. Neural machine translation by jointly learning to align and translate. In *3rd International Conference on Learning Representations, ICLR 2015*, 2015.

15. Jesse Vig. A multiscale visualization of attention in the transformer model. In *Proceedings of the 57th Annual Meeting of the Association for Computational Linguistics: System Demonstrations*, pages 37–42, Florence, Italy, July 2019. Association for Computational Linguistics. https://doi.org/10.18653/v1/P19-3007. https://aclanthology.org/P19-3007.

16. Kelvin Xu, Jimmy Ba, Ryan Kiros, Kyunghyun Cho, Aaron Courville, Ruslan Salakhudinov, Rich Zemel, and Yoshua Bengio. Show, attend and tell: Neural image caption generation with visual attention. In *International conference on machine learning*, pages 2048–2057. PMLR, 2015.

17. Moreno I Coco and Frank Keller. Scan patterns predict sentence production in the cross-modal processing of visual scenes. *Cognitive science*, 36(7):1204–1223, 2012.

18. Ashish Vaswani, Noam Shazeer, Niki Parmar, Jakob Uszkoreit, Llion Jones, Aidan N Gomez, Ł ukasz Kaiser, and Illia Polosukhin. Attention is all you need. In I. Guyon, U. Von Luxburg, S. Bengio, H. Wallach, R. Fergus, S. Vishwanathan, and R. Garnett, editors, *Advances in Neural Information Processing Systems*, volume 30. Curran Associates, Inc., 2017. https://proceedings. neurips.cc/paper/2017/file/3f5ee243547dee91fbd053c1c4a845aa-Paper.pdf.

19. Sarah Wiegreffe and Yuval Pinter. Attention is not not explanation. In *Proceedings of the 2019 Conference on Empirical Methods in Natural Language Processing and the 9th International Joint Conference on Natural Language Processing (EMNLP-IJCNLP)*, pages 11–20, Hong Kong, China, November 2019. Association for Computational Linguistics. https://doi.org/10. 18653/v1/D19-1002. https://aclanthology.org/D19-1002.

20. Sofia Serrano and Noah A. Smith. Is attention interpretable? In *Proceedings of the 57th Annual Meeting of the Association for Computational Linguistics*, pages 2931–2951, Florence, Italy, July 2019. Association for Computational Linguistics. https://doi.org/10.18653/v1/P19-1282. https://aclanthology.org/P19 1282.

21. Danish Pruthi, Mansi Gupta, Bhuwan Dhingra, Graham Neubig, and Zachary C. Lipton. Learning to deceive with attention-based explanations. In *Proceedings of the 58th Annual Meeting of the Association for Computational Linguistics*, pages 4782–4793, Online, July 2020. Association for Computational Linguistics. https://doi.org/10.18653/v1/2020.acl-main.432. https:// aclanthology.org/2020.acl-main.432.

22. Jasmijn Bastings and Katja Filippova. The elephant in the interpretability room: Why use attention as explanation when we have saliency methods? In *Proceedings of the Third BlackboxNLP Workshop on Analyzing and Interpreting Neural Networks for NLP*, pages 149–155, Online, November 2020. Association for Computational Linguistics. https://doi.org/10.18653/v1/2020. blackboxnlp-1.14. https://aclanthology.org/2020.blackboxnlp-1.14.

23. Sarthak Jain and Byron C. Wallace. Attention is not Explanation. In *Proceedings of the 2019 Conference of the North American Chapter of the Association for Computational Linguistics: Human Language Technologies, Volume 1 (Long and Short Papers)*, pages 3543–3556, Minneapolis, Minnesota, June 2019. Association for Computational Linguistics. https://doi.org/10. 18653/v1/N19-1357. https://aclanthology.org/N19-1357.

24. Clara Meister, Stefan Lazov, Isabelle Augenstein, and Ryan Cotterell. Is sparse attention more interpretable? In *Proceedings of the 59th Annual Meeting of the Association for Computational Linguistics and the 11th International Joint Conference on Natural Language Processing (Volume 2: Short Papers)*, pages 122–129, Online, August 2021. Association for Computational Linguistics. https://doi.org/10.18653/v1/2021.acl-short.17. https://aclanthology.org/2021.acl-short.17.

25. Shikhar Vashishth, Shyam Upadhyay, Gaurav Singh Tomar, and Manaal Faruqui. Attention interpretability across nlp tasks. *arXiv preprint* arXiv:1909.11218, 2019.

26. Goro Kobayashi, Tatsuki Kuribayashi, Sho Yokoi, and Kentaro Inui. Attention is not only a weight: Analyzing transformers with vector norms. In *Proceedings of the 2020 Conference on Empirical Methods in Natural Language Processing (EMNLP)*, pages 7057–7075, Online, November 2020. Association for Computational Linguistics. https://doi.org/10.18653/v1/2020. emnlp-main.574. https://aclanthology.org/2020.emnlp-main.574.

27. Samira Abnar and Willem Zuidema. Quantifying attention flow in transformers. In *Proceedings of the 58th Annual Meeting of the Association for Computational Linguistics*, pages 4190–4197, Online, July 2020. Association for Computational Linguistics. https://doi.org/10.18653/ v1/2020.acl-main.385. https://aclanthology.org/2020.acl-main.385.

28. Kawin Ethayarajh and Dan Jurafsky. Attention flows are shapley value explanations. In *Proceedings of the 59th Annual Meeting of the Association for Computational Linguistics and the 11th International Joint Conference on Natural Language Processing (Volume 2: Short Papers)*, pages 49–54, Online, August 2021. Association for Computational Linguistics. https://doi.org/10.18653/v1/2021.acl-short.8. https://aclanthology.org/2021.acl-short.8.

29. Zachary C. Lipton. The mythos of model interpretability: In machine learning, the concept of interpretability is both important and slippery. *Queue*, 16(3):31–57, June 2018. ISSN 1542-7730. https://doi.org/10.1145/3236386.3241340.

30. Karen Simonyan, Andrea Vedaldi, and Andrew Zisserman. Deep inside convolutional networks: Visualising image classification models and saliency maps. In Yoshua Bengio and Yann LeCun, editors, *2nd International Conference on Learning Representations, ICLR 2014, Banff, AB, Canada, April 14-16, 2014, Workshop Track Proceedings*, 2014. http://arxiv.org/abs/1312.6034.

31. Jiwei Li, Xinlei Chen, Eduard Hovy, and Dan Jurafsky. Visualizing and understanding neural models in NLP. In *Proceedings of the 2016 Conference of the North American Chapter of the Association for Computational Linguistics: Human Language Technologies*, pages 681–691, San Diego, California, June 2016. Association for Computational Linguistics. https://doi.org/10.18653/v1/N16-1082. https://aclanthology.org/N16-1082.

32. Marco Ancona, Enea Ceolini, Cengiz Öztireli, and Markus Gross. Gradient-based attribution methods. In *Explainable AI: Interpreting, Explaining and Visualizing Deep Learning*, pages 169–191. Springer, 2019.

33. Avanti Shrikumar, Peyton Greenside, and Anshul Kundaje. Learning important features through propagating activation differences. In *Proceedings of the 34th International Conference on Machine Learning - Volume 70*, ICML'17, page 3145–3153. JMLR.org, 2017.

34. Andreas Madsen. Visualizing memorization in rnns. *Distill*, 2019. https://doi.org/10.23915/distill.00016. https://distill.pub/2019/memorization-in-rnns.

35. Mukund Sundararajan, Ankur Taly, and Qiqi Yan. Axiomatic attribution for deep networks. In Doina Precup and Yee Whye Teh, editors, *Proceedings of the 34th International Conference on Machine Learning*, volume 70 of *Proceedings of Machine Learning Research*, pages 3319–3328. PMLR, 06–11 Aug 2017. https://proceedings.mlr.press/v70/sundararajan17a.html.

36. Alon Jacovi, Swabha Swayamdipta, Shauli Ravfogel, Yanai Elazar, Yejin Choi, and Yoav Goldberg. Contrastive explanations for model interpretability. In *Proceedings of the 2021 Conference on Empirical Methods in Natural Language Processing*, pages 1597–1611, Online and Punta Cana, Dominican Republic, November 2021. Association for Computational Linguistics. https://doi.org/10.18653/v1/2021.emnlp-main.120. https://aclanthology.org/2021.emnlp-main.120.

37. Joseph D Janizek, Pascal Sturmfels, and Su-In Lee. Explaining explanations: Axiomatic feature interactions for deep networks. *J. Mach. Learn. Res.*, 22:104–1, 2021.

38. Sandipan Sikdar, Parantapa Bhattacharya, and Kieran Heese. Integrated directional gradients: Feature interaction attribution for neural NLP models. In *Proceedings of the 59th Annual Meeting of the Association for Computational Linguistics and the 11th International Joint Conference on Natural Language Processing (Volume 1: Long Papers)*, pages 865–878, Online, August 2021. Association for Computational Linguistics. https://doi.org/10.18653/v1/2021.acl-long.71. https://aclanthology.org/2021.acl-long.71.

39. Ian Goodfellow, Yoshua Bengio, and Aaron Courville. *Deep Learning*. MIT Press, 2016. http://www.deeplearningbook.org.

40. Antonio Laverghetta Jr., Animesh Nighojkar, Jamshidbek Mirzakhalov, and John Licato. Can transformer language models predict psychometric properties? In *Proceedings of *SEM 2021: The Tenth Joint Conference on Lexical and Computational Semantics*, pages 12–25, Online,

August 2021. Association for Computational Linguistics. https://doi.org/10.18653/v1/2021. starsem-1.2. https://aclanthology.org/2021.starsem-1.2.

41. Hazel Blythe and Holly Joseph. *Children's Eye Movements during Reading*. 01 2011. https://doi.org/10.1093/oxfordhb/9780199539789.013.0036.

42. Sascha Schroeder, Jukka Hyönä, and Simon Liversedge. Developmental eye-tracking research in reading: Introduction to the special issue. *Journal of Cognitive Psychology*, 27:500–510, 07 2015. https://doi.org/10.1080/20445911.2015.1046877.

43. Michael Hahn and Frank Keller. Modeling human reading with neural attention. In *Proceedings of the 2016 Conference on Empirical Methods in Natural Language Processing*, pages 85–95, Austin, Texas, November 2016. Association for Computational Linguistics. https://doi.org/10.18653/v1/D16-1009. https://aclanthology.org/D16-1009.

44. Shiyang Su and Mark L Davison. Improving the predictive validity of reading comprehension using response times of correct item responses. *Applied Measurement in Education*, 32(2):166–182, 2019.

45. Tal Linzen and T Florian Jaeger. Uncertainty and expectation in sentence processing: Evidence from subcategorization distributions. *Cognitive science*, 40(6):1382–1411, 2016.

46. Stefan L Frank, Leun J Otten, Giulia Galli, and Gabriella Vigliocco. The ERP response to the amount of information conveyed by words in sentences. *Brain and language*, 140:1–11, 2015.

47. Shravan Vasishth, Titus von der Malsburg, and Felix Engelmann. What eye movements can tell us about sentence comprehension. *Wiley Interdisciplinary Reviews: Cognitive Science*, 4(2):125–134, 2013.

48. Jonathan Malmaud, Roger Levy, and Yevgeni Berzak. Bridging information-seeking human gaze and machine reading comprehension. In *Proceedings of the 24th Conference on Computational Natural Language Learning*, pages 142–152, Online, November 2020. Association for Computational Linguistics. https://doi.org/10.18653/v1/2020.conll-1.11. https://aclanthology.org/2020.conll-1.11.

49. Ekta Sood, Simon Tannert, Diego Frassinelli, Andreas Bulling, and Ngoc Thang Vu. Interpreting attention models with human visual attention in machine reading comprehension. In *Proceedings of the 24th Conference on Computational Natural Language Learning*, pages 12–25, Online, November 2020. Association for Computational Linguistics. https://doi.org/10.18653/v1/2020.conll-1.2. https://aclanthology.org/2020.conll-1.2.

50. Ian Tenney, Dipanjan Das, and Ellie Pavlick. BERT rediscovers the classical NLP pipeline. In *Proceedings of the 57th Annual Meeting of the Association for Computational Linguistics*, pages 4593–4601, Florence, Italy, July 2019. Association for Computational Linguistics. https://doi.org/10.18653/v1/P19-1452. https://aclanthology.org/P19-1452.

51. Joshua Bensemann, Alex Peng, Diana Benavides-Prado, Yang Chen, Neset Tan, Paul Michael Corballis, Patricia Riddle, and Michael Witbrock. Eye gaze and self-attention: How humans and transformers attend words in sentences. In *Proceedings of the Workshop on Cognitive Modeling and Computational Linguistics*, pages 75–87, Dublin, Ireland, May 2022. Association for Computational Linguistics. https://doi.org/10.18653/v1/2022.cmcl-1.9. https://aclanthology.org/2022.cmcl-1.9.

52. Oliver Eberle, Stephanie Brandl, Jonas Pilot, and Anders Søgaard. Do transformer models show similar attention patterns to task-specific human gaze? In *Proceedings of the 60th Annual Meeting of the Association for Computational Linguistics (Volume 1: Long Papers)*, pages 4295–4309, Dublin, Ireland, May 2022. Association for Computational Linguistics. https://doi.org/10.18653/v1/2022.acl-long.296. https://aclanthology.org/2022.acl-long.296.

53. Alon Jacovi and Yoav Goldberg. Towards faithfully interpretable NLP systems: How should we define and evaluate faithfulness? In *Proceedings of the 58th Annual Meeting of the Association*

for Computational Linguistics, pages 4198–4205, Online, July 2020. Association for Computational Linguistics. https://doi.org/10.18653/v1/2020.acl-main.386. https://aclanthology.org/2020.acl-main.386.

54. Reinhold Kliegl, Ellen Grabner, Martin Rolfs, and Ralf Engbert. Length, frequency, and predictability effects of words on eye movements in reading. *European Journal of Cognitive Psychology*, 16(1-2):262–284, 2004. https://doi.org/10.1080/09541440340000213. https://doi.org/10.1080/09541440340000213.

55. Simon P Liversedge, Denis Drieghe, Xin Li, Guoli Yan, Xuejun Bai, and Jukka Hyönä. Universality in eye movements and reading: A trilingual investigation. *Cognition*, 147:1–20, 2016.

56. Maryam AlJassmi, Kayleigh Warrington, Victoria McGowan, Sarah White, and Kevin Paterson. Effects of word predictability on eye movements during Arabic reading. *Attention, Perception, & Psychophysics*, 84(1):10–24, 2022. https://doi.org/10.3758/s13414-021-02375-1.

57. Anna K Laurinavichyute, Irina A Sekerina, Svetlana Alexeeva, Kristine Bagdasaryan, and Reinhold Kliegl. Russian sentence corpus: Benchmark measures of eye movements in reading in russian. *Behavior research methods*, 51(3):1161–1178, 2019.

58. Charlotte Pouw, Nora Hollenstein, and Lisa Beinborn. Cross-lingual transfer of cognitive processing complexity. In *Findings of the Association for Computational Linguistics: EACL 2023*, pages 643–657, Dubrovnik, Croatia, May 2023. Association for Computational Linguistics. https://aclanthology.org/2023.findings-eacl.49.

59. Felix Morger, Stephanie Brandl, Lisa Beinborn, and Nora Hollenstein. A cross-lingual comparison of human and model relative word importance. In *Proceedings of the 2022 CLASP Conference on (Dis)embodiment*, pages 11–23, Gothenburg, Sweden, September 2022. Association for Computational Linguistics. https://aclanthology.org/2022.clasp-1.2.

60. Stephanie Brandl and Nora Hollenstein. Every word counts: A multilingual analysis of individual human alignment with model attention. In *Proceedings of the 2nd Conference of the Asia-Pacific Chapter of the Association for Computational Linguistics and the 12th International Joint Conference on Natural Language Processing (Volume 2: Short Papers)*, pages 72–77, Online only, November 2022. Association for Computational Linguistics. https://aclanthology.org/2022.aacl-short.10.

61. Nora Hollenstein, Federico Pirovano, Ce Zhang, Lena Jäger, and Lisa Beinborn. Multilingual language models predict human reading behavior. In *Proceedings of the 2021 Conference of the North American Chapter of the Association for Computational Linguistics: Human Language Technologies*, pages 106–123, Online, June 2021. Association for Computational Linguistics. https://doi.org/10.18653/v1/2021.naacl-main.10. https://aclanthology.org/2021.naacl-main.10.

62. Nora Hollenstein, Emmanuele Chersoni, Cassandra L. Jacobs, Yohei Oseki, Laurent Prévot, and Enrico Santus. CMCL 2021 shared task on eye-tracking prediction. In *Proceedings of the Workshop on Cognitive Modeling and Computational Linguistics*, pages 72–78, Online, June 2021. Association for Computational Linguistics. https://doi.org/10.18653/v1/2021.cmcl-1.7. https://aclanthology.org/2021.cmcl-1.7.

63. Nora Hollenstein, Emmanuele Chersoni, Cassandra Jacobs, Yohei Oseki, Laurent Prévot, and Enrico Santus. CMCL 2022 shared task on multilingual and crosslingual prediction of human reading behavior. In *Proceedings of the Workshop on Cognitive Modeling and Computational Linguistics*, pages 121–129, Dublin, Ireland, May 2022. Association for Computational Linguistics. https://doi.org/10.18653/v1/2022.cmcl-1.14. https://aclanthology.org/2022.cmcl-1.14.

64. Sunit Bhattacharya, Rishu Kumar, and Ondrej Bojar. Team ÚFAL at CMCL 2022 shared task: Figuring out the correct recipe for predicting eye-tracking features using pretrained language models. In *Proceedings of the Workshop on Cognitive Modeling and Computational Linguistics*,

pages 130–135, Dublin, Ireland, May 2022. Association for Computational Linguistics. https://doi.org/10.18653/v1/2022.cmcl-1.15. https://aclanthology.org/2022.cmcl-1.15.

65. Ece Takmaz. Team DMG at CMCL 2022 shared task: Transformer adapters for the multi- and cross-lingual prediction of human reading behavior. In *Proceedings of the Workshop on Cognitive Modeling and Computational Linguistics*, pages 136–144, Dublin, Ireland, May 2022. Association for Computational Linguistics. https://doi.org/10.18653/v1/2022.cmcl-1.16. https://aclanthology.org/2022.cmcl-1.16.

66. Harshvardhan Srivastava. Poirot at CMCL 2022 shared task: Zero shot crosslingual eye-tracking data prediction using multilingual transformer models. In *Proceedings of the Workshop on Cognitive Modeling and Computational Linguistics*, pages 102–107, Dublin, Ireland, May 2022. Association for Computational Linguistics. https://doi.org/10.18653/v1/2022.cmcl-1.11. https://aclanthology.org/2022.cmcl-1.11.

67. Daniel Wiechmann, Yu Qiao, Elma Kerz, and Justus Mattern. Measuring the impact of (psycho-)linguistic and readability features and their spill over effects on the prediction of eye move-ment patterns. In *Proceedings of the 60th Annual Meeting of the Association for Computational Linguistics (Volume 1: Long Papers)*, pages 5276–5290, Dublin, Ireland, May 2022. Asso-ciation for Computational Linguistics. https://doi.org/10.18653/v1/2022.acl-long.362. https://aclanthology.org/2022.acl-long.362.

68. Lyn Frazier. *On comprehending sentences: Syntactic parsing strategies*. University of Con-necticut, 1979.

69. John Hale. A probabilistic Earley parser as a psycholinguistic model. In *Second Meeting of the North American Chapter of the Association for Computational Linguistics*, 2001. https://aclanthology.org/N01-1021.

70. Nathaniel J. Smith and Roger Levy. The effect of word predictability on reading time is logarith-mic. *Cognition*, 128(3):302–319, 2013. ISSN 0010-0277. https://doi.org/10.1016/j.cognition.2013.02.013. https://www.sciencedirect.com/science/article/pii/S0010027713000413.

71. Richard Futrell and Roger Levy. Noisy-context surprisal as a human sentence processing cost model. In *Proceedings of the 15th Conference of the European Chapter of the Association for Computational Linguistics: Volume 1, Long Papers*, pages 688–698, Valencia, Spain, April 2017. Association for Computational Linguistics. https://aclanthology.org/E17-1065.

72. Cory Shain, Marten van Schijndel, Richard Futrell, Edward Gibson, and William Schuler. Memory access during incremental sentence processing causes reading time latency. In *Pro-ceedings of the Workshop on Computational Linguistics for Linguistic Complexity (CL4LC)*, pages 49–58, Osaka, Japan, December 2016. The COLING 2016 Organizing Committee. https://aclanthology.org/W16-4106.

73. Edward Gibson et al. The dependency locality theory: A distance-based theory of linguistic complexity. *Image, language, brain*, 2000:95–126, 2000.

74. Edward Gibson and James Thomas. Memory limitations and structural forgetting: The per-ception of complex ungrammatical sentences as grammatical. *Language and Cognitive Pro-cesses*, 14(3):225–248, 1999. https://doi.org/10.1080/016909699386293. https://doi.org/10.1080/016909699386293.

75. Shravan Vasishth, Katja Suckow, Richard L. Lewis, and Sabine Kern. Short-term forgetting in sentence comprehension: Crosslinguistic evidence from verb-final structures. *Language and Cognitive Processes*, 25(4):533–567, 2010. https://doi.org/10.1080/01690960903310587. https://doi.org/10.1080/01690960903310587.

76. Richard Futrell, Ethan Wilcox, Takashi Morita, Peng Qian, Miguel Ballesteros, and Roger Levy. Neural language models as psycholinguistic subjects: Representations of syntactic state. In *Proceedings of the 2019 Conference of the North American Chapter of the Association for Computational Linguistics: Human Language Technologies, Volume 1 (Long and Short Papers)*,

        pages 32–42, Minneapolis, Minnesota, June 2019. Association for Computational Linguistics.
        https://doi.org/10.18653/v1/N19-1004. https://aclanthology.org/N19-1004.

77. Stefan L Frank, Thijs Trompenaars, and Shravan Vasishth. Cross-linguistic differences in pro-
    cessing double-embedded relative clauses: Working-memory constraints or language statistics?
    Cognitive Science, 40(3):554–578, 2016.

78. Daniel Gildea and T Florian Jaeger. Human languages order information efficiently. *arXiv
    preprint* arXiv:1510.02823, 2015.

79. Edward Gibson, Richard Futrell, Steven P. Piantadosi, Isabelle Dautriche, Kyle Mahowald,
    Leon Bergen, and Roger Levy. How efficiency shapes human language. *Trends in Cognitive
    Sciences*, 23(5):389–407, 2019. ISSN 1364-6613. https://doi.org/10.1016/j.tics.2019.02.003.
    https://www.sciencedirect.com/science/article/pii/S1364661319300580.

80. Stefan L Frank and Roel M Willems. Word predictability and semantic similarity show distinct
    patterns of brain activity during language comprehension. *Language, Cognition and Neuro-
    science*, 32(9):1192–1203, 2017.

81. Sharmistha Jat, Hao Tang, Partha Talukdar, and Tom Mitchell. Relating simple sentence
    representations in deep neural networks and the brain. In *Proceedings of the 57th Annual
    Meeting of the Association for Computational Linguistics*, pages 5137–5154, Florence, Italy,
    July 2019. Association for Computational Linguistics. https://doi.org/10.18653/v1/P19-1507.
    https://aclanthology.org/P19-1507.

82. Danny Merkx and Stefan L. Frank. Human sentence processing: Recurrence or attention? In
    Proceedings of the Workshop on Cognitive Modeling and Computational Linguistics, pages 12–
    22, Online, June 2021. Association for Computational Linguistics. https://doi.org/10.18653/v1/
    2021.cmcl-1.2. https://aclanthology.org/2021.cmcl-1.2.

83. James A Michaelov, Megan D Bardolph, Seana Coulson, and Benjamin K Bergen. Differ-
    ent kinds of cognitive plausibility: why are transformers better than rnns at predicting n400
    amplitude? *arXiv preprint* arXiv:2107.09648, 2021.

84. Ethan Gotlieb Wilcox, Jon Gauthier, Jennifer Hu, Peng Qian, and Roger Levy. On the predictive
    power of neural language models for human real-time comprehension behavior. *arXiv preprint*
    arXiv:2006.01912, 2020.

85. Tiwalayo Eisape, Noga Zaslavsky, and Roger Levy. Cloze distillation improves psychometric
    predictive power. In *Proceedings of the 24th Conference on Computational Natural Language
    Learning*, pages 609–619, 2020.

86. Martin Schrimpf, Idan Blank, Greta Tuckute, Carina Kauf, Eghbal A Hosseini, Nancy Kan-
    wisher, Joshua Tenenbaum, and Evelina Fedorenko. Artificial neural networks accurately pre-
    dict language processing in the brain. *BioRxiv*, 10(2020.06):26–174482, 2020.

87. Matthew J Nelson, Imen El Karoui, Kristof Giber, Xiaofang Yang, Laurent Cohen, Hilda Koop-
    man, Sydney S Cash, Lionel Naccache, John T Hale, Christophe Pallier, et al. Neurophysio-
    logical dynamics of phrase-structure building during sentence processing. *Proceedings of the
    National Academy of Sciences*, 114(18):E3669–E3678, 2017.

88. Tal Linzen and Marco Baroni. Syntactic structure from deep learning. *Annual Review of Lin-
    guistics*, 7:195–212, 2021.

89. John Hale, Chris Dyer, Adhiguna Kuncoro, and Jonathan Brennan. Finding syntax in human
    encephalography with beam search. In *Proceedings of the 56th Annual Meeting of the Associ-
    ation for Computational Linguistics (Volume 1: Long Papers)*, pages 2727–2736, Melbourne,
    Australia, July 2018. Association for Computational Linguistics. https://doi.org/10.18653/v1/
    P18-1254. https://aclanthology.org/P18-1254.

90. R. Thomas McCoy, Robert Frank, and Tal Linzen. Does syntax need to grow on trees? sources of
    hierarchical inductive bias in sequence-to-sequence networks. *Transactions of the Association*

for Computational Linguistics, 8:125–140, 2020. https://doi.org/10.1162/tacl_a_00304. https://aclanthology.org/2020.tacl-1.9.

91. Timothy P McNamara. Priming and constraints it places on theories of memory and retrieval. *Psychological Review*, 99(4):650, 1992.

92. Kanishka Misra, Allyson Ettinger, and Julia Rayz. Exploring BERT's sensitivity to lexical cues using tests from semantic priming. In *Findings of the Association for Computational Linguistics: EMNLP 2020*, pages 4625–4635, Online, November 2020. Association for Computational Linguistics. https://doi.org/10.18653/v1/2020.findings-emnlp.415. https://aclanthology.org/2020.findings-emnlp.415.

93. Kristen M Tooley and Matthew J Traxler. Syntactic priming effects in comprehension: A critical review. *Language and Linguistics Compass*, 4(10):925–937, 2010.

94. Grusha Prasad, Marten van Schijndel, and Tal Linzen. Using priming to uncover the organization of syntactic representations in neural language models. In *Proceedings of the 23rd Conference on Computational Natural Language Learning (CoNLL)*, pages 66–76, Hong Kong, China, November 2019. Association for Computational Linguistics. https://doi.org/10.18653/v1/K19-1007. https://aclanthology.org/K19-1007.

95. Alona Fyshe, Partha P. Talukdar, Brian Murphy, and Tom M. Mitchell. Interpretable semantic vectors from a joint model of brain- and text- based meaning. In *Proceedings of the 52nd Annual Meeting of the Association for Computational Linguistics (Volume 1: Long Papers)*, pages 489–499, Baltimore, Maryland, June 2014. Association for Computational Linguistics. https://doi.org/10.3115/v1/P14-1046. https://aclanthology.org/P14-1046.

96. Dan Schwartz, Mariya Toneva, and Leila Wehbe. Inducing brain-relevant bias in natural language processing models. *Advances in neural information processing systems*, 32, 2019.

97. Alex Wang, Amanpreet Singh, Julian Michael, Felix Hill, Omer Levy, and Samuel Bowman. GLUE: A multi-task benchmark and analysis platform for natural language understanding. In *Proceedings of the 2018 EMNLP Workshop BlackboxNLP: Analyzing and Interpreting Neural Networks for NLP*, pages 353–355, Brussels, Belgium, November 2018. Association for Computational Linguistics. https://doi.org/10.18653/v1/W18-5446. https://aclanthology.org/W18-5446.

98. Graham Kalton and Howard Schuman. The effect of the question on survey responses: A review. *Journal of the Royal Statistical Society: Series A (General)*, 145(1):42–57, 1982.

99. Nora Kassner and Hinrich Schütze. Negated and misprimed probes for pretrained language models: Birds can talk, but cannot fly. In *Proceedings of the 58th Annual Meeting of the Association for Computational Linguistics*, pages 7811–7818, Online, July 2020. Association for Computational Linguistics. https://doi.org/10.18653/v1/2020.acl-main.698. https://aclanthology.org/2020.acl-main.698.

100. Wai Man Si, Michael Backes, Jeremy Blackburn, Emiliano De Cristofaro, Gianluca Stringhini, Savvas Zannettou, and Yang Zhang. Why so toxic? measuring and triggering toxic behavior in open-domain chatbots. In *Proceedings of the 2022 ACM SIGSAC Conference on Computer and Communications Security*, CCS '22, page 2659–2673, New York, NY, USA, 2022. Association for Computing Machinery. ISBN 9781450394505. https://doi.org/10.1145/3548606.3560599. https://doi.org/10.1145/3548606.3560599.

101. Matt Gardner, Yoav Artzi, Victoria Basmov, Jonathan Berant, Ben Bogin, Sihao Chen, Pradeep Dasigi, Dheeru Dua, Yanai Elazar, Ananth Gottumukkala, Nitish Gupta, Hannaneh Hajishirzi, Gabriel Ilharco, Daniel Khashabi, Kevin Lin, Jiangming Liu, Nelson F. Liu, Phoebe Mulcaire, Qiang Ning, Sameer Singh, Noah A. Smith, Sanjay Subramanian, Reut Tsarfaty, Eric Wallace, Ally Zhang, and Ben Zhou. Evaluating models' local decision boundaries via contrast sets. In *Findings of the Association for Computational Linguistics: EMNLP 2020*, pages 1307–1323,

Online, November 2020. Association for Computational Linguistics. https://doi.org/10.18653/v1/2020.findings-emnlp.117. https://aclanthology.org/2020.findings-emnlp.117.

102. Rich Caruana. Multitask learning. *Machine learning*, 28(1):41–75, 1997.

103. Sebastian Ruder. An overview of multi-task learning in deep neural networks. http://ruder.io/multi-task, 2017.

104. Ronan Collobert, Jason Weston, Léon Bottou, Michael Karlen, Koray Kavukcuoglu, and Pavel Kuksa. Natural language processing (almost) from scratch. *Journal of machine learning research*, 12(ARTICLE):2493–2537, 2011.

105. Nora Hollenstein, Maria Barrett, Marius Troendle, Francesco Bigiolli, Nicolas Langer, and Ce Zhang. Advancing nlp with cognitive language processing signals. *arXiv preprint* arXiv:1904.02682, 2019.

106. Ana Valeria González-Garduño and Anders Søgaard. Using gaze to predict text readability. In *Proceedings of the 12th Workshop on Innovative Use of NLP for Building Educational Applications*, pages 438–443, Copenhagen, Denmark, September 2017. Association for Computational Linguistics. https://doi.org/10.18653/v1/W17-5050. https://aclanthology.org/W17-5050.

107. Sigrid Klerke, Yoav Goldberg, and Anders Søgaard. Improving sentence compression by learning to predict gaze. In *Proceedings of the 2016 Conference of the North American Chapter of the Association for Computational Linguistics: Human Language Technologies*, pages 1528–1533, San Diego, California, June 2016. Association for Computational Linguistics. https://doi.org/10.18653/v1/N16-1179. https://aclanthology.org/N16-1179.

108. Sigrid Klerke and Barbara Plank. At a glance: The impact of gaze aggregation views on syntactic tagging. In *Proceedings of the Beyond Vision and LANguage: inTEgrating Real-world kNowledge (LANTERN)*, pages 51–61, Hong Kong, China, November 2019. Association for Computational Linguistics. https://doi.org/10.18653/v1/D19-6408. https://aclanthology.org/D19-6408.

109. Erik McGuire and Noriko Tomuro. Relation classification with cognitive attention supervision. In *Proceedings of the Workshop on Cognitive Modeling and Computational Linguistics*, pages 222–232, Online, June 2021. Association for Computational Linguistics. https://doi.org/10.18653/v1/2021.cmcl-1.26. https://aclanthology.org/2021.cmcl-1.26.

110. Maria Barrett, Joachim Bingel, Nora Hollenstein, Marek Rei, and Anders Søgaard. Sequence classification with human attention. In *Proceedings of the 22nd Conference on Computational Natural Language Learning*, pages 302–312, Brussels, Belgium, October 2018. Association for Computational Linguistics. https://doi.org/10.18653/v1/K18-1030. https://aclanthology.org/K18-1030.

111. Lukas Muttenthaler, Nora Hollenstein, and Maria Barrett. Human brain activity for machine attention. *arXiv preprint* arXiv:2006.05113, 2020.

112. Sidney Evaldo Leal, João Marcos Munguba Vieira, Erica dos Santos Rodrigues, Elisângela Nogueira Teixeira, and Sandra Aluísio. Using eye-tracking data to predict the readability of Brazilian Portuguese sentences in single-task, multi-task and sequential transfer learning approaches. In *Proceedings of the 28th International Conference on Computational Linguistics*, pages 5821–5831, Barcelona, Spain (Online), December 2020. International Committee on Computational Linguistics. https://doi.org/10.18653/v1/2020.coling-main.512. https://aclanthology.org/2020.coling-main.512.

113. Mariya Toneva and Leila Wehbe. Interpreting and improving natural-language processing (in machines) with natural language-processing (in the brain). *Advances in Neural Information Processing Systems*, 32, 2019.

114. Yifei Luo, Minghui Xu, and Deyi Xiong. CogTaskonomy: Cognitively inspired task taxonomy is beneficial to transfer learning in NLP. In *Proceedings of the 60th Annual Meeting of the Association for Computational Linguistics (Volume 1: Long Papers)*, pages 904–920, Dublin,

Ireland, May 2022. Association for Computational Linguistics. https://doi.org/10.18653/v1/2022.acl-long.64. https://aclanthology.org/2022.acl-long.64.

115. James Kirkpatrick, Razvan Pascanu, Neil Rabinowitz, Joel Veness, Guillaume Desjardins, Andrei A Rusu, Kieran Milan, John Quan, Tiago Ramalho, Agnieszka Grabska-Barwinska, et al. Overcoming catastrophic forgetting in neural networks. *Proceedings of the national academy of sciences*, 114(13):3521–3526, 2017.

116. Christopher D Manning, Kevin Clark, John Hewitt, Urvashi Khandelwal, and Omer Levy. Emergent linguistic structure in artificial neural networks trained by self-supervision. *Proceedings of the National Academy of Sciences*, 117(48):30046–30054, 2020.

117. Alex Warstadt and Samuel R Bowman. Can neural networks acquire a structural bias from raw linguistic data? In *CogSci*, 2020.

118. Alex Warstadt, Yian Zhang, Xiaocheng Li, Haokun Liu, and Samuel R. Bowman. Learning which features matter: RoBERTa acquires a preference for linguistic generalizations (eventually). In *Proceedings of the 2020 Conference on Empirical Methods in Natural Language Processing (EMNLP)*, pages 217–235, Online, November 2020. Association for Computational Linguistics. https://doi.org/10.18653/v1/2020.emnlp-main.16. https://aclanthology.org/2020.emnlp-main.16.

119. Marten van Schijndel, Aaron Mueller, and Tal Linzen. Quantity doesn't buy quality syntax with neural language models. In *Proceedings of the 2019 Conference on Empirical Methods in Natural Language Processing and the 9th International Joint Conference on Natural Language Processing (EMNLP-IJCNLP)*, pages 5831–5837, Hong Kong, China, November 2019. Association for Computational Linguistics. https://doi.org/10.18653/v1/D19-1592. https://aclanthology.org/D19-1592.

120. Nicholas Carlini, Daphne Ippolito, Matthew Jagielski, Katherine Lee, Florian Tramer, and Chiyuan Zhang. Quantifying memorization across neural language models. In *The Eleventh International Conference on Learning Representations*, 2023. https://openreview.net/forum?id=TatRHT_1cK.

121. Chris Dyer, Adhiguna Kuncoro, Miguel Ballesteros, and Noah A. Smith. Recurrent neural network grammars. In *Proceedings of the 2016 Conference of the North American Chapter of the Association for Computational Linguistics: Human Language Technologies*, pages 199–209, San Diego, California, June 2016. Association for Computational Linguistics. https://doi.org/10.18653/v1/N16-1024. https://aclanthology.org/N16-1024.

122. Mostafa Abdou, Ana Valeria González, Mariya Toneva, Daniel Hershcovich, and Anders Søgaard. Does injecting linguistic structure into language models lead to better alignment with brain recordings? *arXiv preprint* arXiv:2101.12608, 2021.

123. T Florian Jaeger. Redundancy and reduction: Speakers manage syntactic information density. *Cognitive psychology*, 61(1):23–62, 2010.

124. Jason Wei, Clara Meister, and Ryan Cotterell. A cognitive regularizer for language modeling. In *Proceedings of the 59th Annual Meeting of the Association for Computational Linguistics and the 11th International Joint Conference on Natural Language Processing (Volume 1: Long Papers)*, pages 5191–5202, Online, August 2021. Association for Computational Linguistics. https://doi.org/10.18653/v1/2021.acl-long.404. https://aclanthology.org/2021.acl-long.404.

125. Luke Maurits, Dan Navarro, and Amy Perfors. Why are some word orders more common than others? a uniform information density account. *Advances in neural information processing systems*, 23, 2010.

126. Tatsuki Kuribayashi, Yohei Oseki, Takumi Ito, Ryo Yoshida, Masayuki Asahara, and Kentaro Inui. Lower perplexity is not always human-like. In *Proceedings of the 59th Annual Meeting of the Association for Computational Linguistics and the 11th International Joint Conference on Natural Language Processing (Volume 1: Long Papers)*, pages 5203–5217, Online, August

2021. Association for Computational Linguistics. https://doi.org/10.18653/v1/2021.acl-long. 405. https://aclanthology.org/2021.acl-long.405.

127. Byung-Doh Oh and William Schuler. Why does surprisal from larger transformer-based language models provide a poorer fit to human reading times? *Transactions of the Association for Computational Linguistics*, 11:336–350, 2023.

128. Marten van Schijndel and Tal Linzen. Can entropy explain successor surprisal effects in reading? In *Proceedings of the Society for Computation in Linguistics (SCiL) 2019*, pages 1–7, 2019. https://doi.org/10.7275/qtbb-9d05. https://aclanthology.org/W19-0101.

129. Suhas Arehalli, Brian Dillon, and Tal Linzen. Syntactic surprisal from neural models predicts, but underestimates, human processing difficulty from syntactic ambiguities. In *Proceedings of the 26th Conference on Computational Natural Language Learning (CoNLL)*, pages 301–313, Abu Dhabi, United Arab Emirates (Hybrid), December 2022. Association for Computational Linguistics. https://aclanthology.org/2022.conll-1.20.

130. Ethan Wilcox, Pranali Vani, and Roger Levy. A targeted assessment of incremental processing in neural language models and humans. In *Proceedings of the 59th Annual Meeting of the Association for Computational Linguistics and the 11th International Joint Conference on Natural Language Processing (Volume 1: Long Papers)*, pages 939–952, Online, August 2021. Association for Computational Linguistics. https://doi.org/10.18653/v1/2021.acl-long.76. https://aclanthology.org/2021.acl-long.76.

131. Clara Meister and Ryan Cotterell. Language model evaluation beyond perplexity. In *Proceedings of the 59th Annual Meeting of the Association for Computational Linguistics and the 11th International Joint Conference on Natural Language Processing (Volume 1: Long Papers)*, pages 5328–5339, Online, August 2021. Association for Computational Linguistics. https://doi.org/10.18653/v1/2021.acl-long.414. https://aclanthology.org/2021.acl-long.414.

132. Forrest Davis and Marten van Schijndel. Uncovering constraint-based behavior in neural models via targeted fine-tuning. In *Proceedings of the 59th Annual Meeting of the Association for Computational Linguistics and the 11th International Joint Conference on Natural Language Processing (Volume 1: Long Papers)*, pages 1159–1171, Online, August 2021. Association for Computational Linguistics. https://doi.org/10.18653/v1/2021.acl-long.93. https://aclanthology.org/2021.acl-long.93.

133. Ninareh Mehrabi, Fred Morstatter, Nripsuta Saxena, Kristina Lerman, and Aram Galstyan. A survey on bias and fairness in machine learning. *ACM Comput. Surv.*, 54(6), jul 2021. ISSN 0360-0300. https://doi.org/10.1145/3457607. https://doi.org/10.1145/3457607.

134. Gina Neff. Talking to bots: Symbiotic agency and the case of tay. *International Journal of Communication*, 2016.

135. Sakiko Fukuda-Parr and Elizabeth Gibbons. Emerging consensus on 'ethical ai': Human rights critique of stakeholder guidelines. *Global Policy*, 12:32–44, 2021.

136. Jason Borenstein and Ayanna Howard. Emerging challenges in ai and the need for ai ethics education. *AI and Ethics*, 1(1):61–65, 2021.

137. Emily M. Bender, Dirk Hovy, and Alexandra Schofield. Integrating ethics into the NLP curriculum. In *Proceedings of the 58th Annual Meeting of the Association for Computational Linguistics: Tutorial Abstracts*, pages 6–9, Online, July 2020. Association for Computational Linguistics. https://doi.org/10.18653/v1/2020.acl-tutorials.2. https://aclanthology.org/2020.acl-tutorials.2.

138. Thilo Hagendorff. The ethics of ai ethics: An evaluation of guidelines. *Minds and Machines*, 30(1):99–120, 2020.

139. Brent Mittelstadt. Principles alone cannot guarantee ethical ai. *Nature Machine Intelligence*, 1(11):501–507, 2019.

140. Merve Hickok. Lessons learned from ai ethics principles for future actions. *AI and Ethics*, 1(1):41–47, 2021.
141. UN System CEB for Coordination. Principles for the Ethical Use of Artificial Intelligence in the United Nations System. 2022. https://unsceb.org/sites/default/files/2022-09/Principles%20for %20the%20Ethical%20Use%20of%20AI%20in%20the%20UN%20System_1.pdf.
142. European Commission. Independent high-level expert group on artificial intelligence: Ethics guidelines for trustworthy ai. *Search in*, 2019. https://www.aepd.es/sites/default/files/2019-12/ ai-ethics-guidelines.pdf.

Towards Cognitively More Plausible Models

When initiating the writing process for this book, we hoped that it would bring us closer to a definition of the properties that make a model cognitively plausible. Even when abstracting away from biological plausibility as we did in this book, we cannot yet identify a single approach that will simultaneously lead to higher cognitive plausibility on all of the three dimensions we discussed: behavioral patterns, representational structure, and procedural strategies. We think that it is more useful to interpret cognitive plausibility as a graded concept and evaluate models comparatively by clearly defining specific aspects of cognitive plausibility. Instead of focusing on models that achieve optimal performance under ideal conditions, we should emphasize aspects of impaired performance under degraded conditions [1].

Cognitive signals are an important source of information about human language processing. As NLP researchers, we need to carefully involve knowledge from psycholinguistics and neuroscience to better understand the level of information that can be extracted from each signal type and their methodological drawbacks (temporal and spatial accuracy, signal-to-noise ratio, etc.). In turn, our experience with data processing and comparative evaluation can lead to improved reproducibility of experiments and more robust modeling choices. We think that more elaborate modeling techniques need to be developed to better account for individual and cross-lingual differences.

On the behavioral level, we identified instance difficulty as an underestimated criterion for evaluating cognitive plausibility. The average performance of a model often gives a misleading impression of its abilities. We think that different error types can be associated with different levels of severity. A model is more cognitively plausible if instance-level evaluation metrics such as instance accuracy [2] and overlap ratios [3] correlate with measures of instance difficulty, such as annotator disagreement, or indications of cognitive load (see Chaps. 3 and 4). Methods for modeling disagreement are closely related to interpretability methods for modeling model uncertainty [4–6]. Instead of only focusing on performance, an analysis of the flaws and failures of language understanding and the types of errors that occur

© The Author(s), under exclusive license to Springer Nature Switzerland AG 2024   153
L. Beinborn and N. Hollenstein, *Cognitive Plausibility in Natural Language Processing*, Synthesis Lectures on Human Language Technologies,
https://doi.org/10.1007/978-3-031-43260-6_7

can provide important insights into the different decision patterns in humans and models. Models that can predict human difficulties (for example, when processing new information) can be useful in educational settings. And the interaction between human users and models will be improved if users can reliably anticipate and correct errors in the model output.

Static representational analyses of concepts, properties, and relations cannot easily be mapped to contextualized language models because of the increasing compositional integration of context over layers and the dynamics of finetuning. Current probing methods derive static representations from contextual models to test for the encoded knowledge (see Chap. 5). More dynamic methods that account for the vertical information flow across layers such as layer-wise relevance propagation [7] and for horizontal procedural strategies such as attention flow [8] need to be further developed to better understand how basic representational units are recombined and adjusted in task-specific scenarios. We think that the integration of educational knowledge in difficulty-based curriculum learning [9] and learner-oriented knowledge tracing [10] open up promising perspectives towards more cognitively plausible learning processes.

From a cognitive perspective, the representational focus on form is too naive. However, current multimodal approaches do not yet succeed in efficiently mediating complementary information from different modalities. We need to learn more about cognitive grounding processes from neuroscientific research and language needs to be interpreted as part of a larger system for understanding and communicating [11]. Poeppel et al. [12] envisioned a mapping of the elementary units of linguistic processing to their neurobiological counterparts that has not yet been realized. According to Hale et al. [13], the integration of linguistically interpretable models will play a pivotal role in facilitating the reciprocal exchange between natural language processing and neuroscience. We agree with Abdou [14] that models need not be inherently interpretable if we can use interpretability methods to understand the underlying representational structure and procedural strategies (see Chap. 6).

In the ethical considerations of this book, we highlighted different forms of biases that play a role in natural language processing research. But not all forms of bias are harmful. The inductive bias of a language model indicates how the model weighs and combines input information to predict tokens [15]. The effect of architectural choices on the inductive bias on a neural language model is not yet properly understood as only a tiny fraction of the hypothesis space can be explored at a time [16]. We sketched approaches to examine procedural strategies of relative importance from a cognitive perspective (see Chap. 6) but this field of research is still in its early stages. We expect that further research on transfer and multi-task learning will lead to more generalizable models if we combine tasks in a cognitively plausible way. This means that we need to integrate domain knowledge about the cognitive processes involved in solving a task to determine more suitable target objectives. In dialogue models, human expertise can be directly integrated in an interactive fashion. Reinforcement learning and imitation learning strategies coupled with human feedback have been found to be a promising approach to adjust the model towards cognitively more plausible conversational strategies, for example, by integrating acknowledgment to the user's turns [17, 18].

Earlier studies discussing the cognitive plausibility of language processing models were closely linked to syntactic parsing [19, 20]. Keller [19] specifies that cognitively plausible models need to be "predictive", "efficient", "robust", and have a "broad coverage". Large neural language models have improved the performance significantly in these respects at the loss of modeling transparency which makes it harder to evaluate whether their procedural strategies reflect "incremental" processing and "distance-based memory costs". Lindes and Laird [20] specified ten properties of cognitively plausible models which we want to briefly re-evaluate. Neural language models implement an "integrated" learning process that extracts syntactic and semantic information jointly from the input. They represent "context-dependent" meaning and facilitate "compositional" interpretation by using sub-word tokenization, layerwise processing, residual connections, and attention mechanisms. The compositional principles are not yet well understood from a cognitive perspective and it remains unclear whether "hierarchical" processing constraints are incorporated but novel interpretability methods are being developed to shed light on these questions [21, 22]. Language models can operate almost in "real time" but they need to be finetuned to be "useful" in the sense that linguistic input is mapped into "actionable intelligence". In order to develop cognitively more plausible models, we need to better understand the role of word order [23–25] to facilitate "incremental" and "repair-based processing" mechanisms in computational models. It still remains a major challenge to develop language processing models that are "grounded" in perception, action, and world knowledge which seems to be a precondition for "eclectic" disambiguation using semantic, pragmatic, and world knowledge.

Keller [19] criticized that psycholinguistic models of language processing have been developed almost exclusively for English. In recent years, efforts towards cross-lingual approaches and the integration of low-resource languages have been increased but a strong dominance of languages spoken in economically privileged countries is still undeniable [26]. We strongly believe that a multilingual point of view is required to better understand the cognitive principles underlying general language processing and not only those pertinent to English. We are optimistic that research efforts toward cognitive plausibility will not only lead to better computational models but will also circle back to improving our understanding of the human brain.

References

1. Valerie G. Hardcastle and Kiah Hardcastle. Marr's levels revisited: Understanding how brains break. *Topics in Cognitive Science*, 7(2):259–273, 2015. https://doi.org/10.1111/tops.12130. https://onlinelibrary.wiley.com/doi/abs/10.1111/tops.12130
2. Ruiqi Zhong, Dhruba Ghosh, Dan Klein, and Jacob Steinhardt. Are larger pretrained language models uniformly better? comparing performance at the instance level. In *Findings of the Association for Computational Linguistics: ACL-IJCNLP 2021*, pages 3813–3827, Online, August 2021. Association for Computational Linguistics. https://doi.org/10.18653/v1/2021.findings-acl.334. https://aclanthology.org/2021.findings-acl.334.

3. Urja Khurana, Eric Nalisnick, and Antske Fokkens. How emotionally stable is ALBERT? testing robustness with stochastic weight averaging on a sentiment analysis task. In *Proceedings of the 2nd Workshop on Evaluation and Comparison of NLP Systems*, pages 16–31, Punta Cana, Dominican Republic, November 2021. Association for Computational Linguistics. https://doi.org/10.18653/v1/2021.eval4nlp-1.3. https://aclanthology.org/2021.eval4nlp-1.3

4. Valerio Basile, Michael Fell, Tommaso Fornaciari, Dirk Hovy, Silviu Paun, Barbara Plank, Massimo Poesio, and Alexandra Uma. We need to consider disagreement in evaluation. In *Proceedings of the 1st Workshop on Benchmarking: Past, Present and Future*, pages 15–21, Online, August 2021. Association for Computational Linguistics. https://doi.org/10.18653/v1/2021.bppf-1.3. https://aclanthology.org/2021.bppf-1.3

5. Ellie Pavlick and Tom Kwiatkowski. Inherent disagreements in human textual inferences. *Transactions of the Association for Computational Linguistics*, 7:677–694, 2019. https://doi.org/10.1162/tacl_a_00293. https://aclanthology.org/Q19-1043

6. Edwin Simpson and Iryna Gurevych. Scalable bayesian preference learning for crowds. *Mach. Learn.*, 109(4):689–718, apr 2020. ISSN 0885-6125. https://doi.org/10.1007/s10994-019-05867-2.

7. Leila Arras, Ahmed Osman, Klaus-Robert Müller, and Wojciech Samek. Evaluating recurrent neural network explanations. In *Proceedings of the 2019 ACL Workshop BlackboxNLP: Analyzing and Interpreting Neural Networks for NLP*, pages 113–126, Florence, Italy, August 2019. Association for Computational Linguistics. https://doi.org/10.18653/v1/W19-4813. https://aclanthology.org/W19-4813.

8. Samira Abnar and Willem Zuidema. Quantifying attention flow in transformers. In *Proceedings of the 58th Annual Meeting of the Association for Computational Linguistics*, pages 4190–4197, Online, July 2020. Association for Computational Linguistics. https://doi.org/10.18653/v1/2020.acl-main.385. https://aclanthology.org/2020.acl-main.385.

9. Petru Soviany, Radu Tudor Ionescu, Paolo Rota, and Nicu Sebe. Curriculum learning: A survey. *International Journal of Computer Vision*, pages 1–40, 2022.

10. Chris Piech, Jonathan Bassen, Jonathan Huang, Surya Ganguli, Mehran Sahami, Leonidas J Guibas, and Jascha Sohl-Dickstein. Deep knowledge tracing. In C. Cortes, N. Lawrence, D. Lee, M. Sugiyama, and R. Garnett, editors, *Advances in Neural Information Processing Systems*, volume 28. Curran Associates, Inc., 2015. https://proceedings.neurips.cc/paper/2015/file/bac9162b47c56fc8a4d2a519803d51b3-Paper.pdf.

11. James L McClelland, Felix Hill, Maja Rudolph, Jason Baldridge, and Hinrich Schütze. Extending machine language models toward human-level language understanding. *arXiv preprint* arXiv:1912.05877, 2019.

12. David Poeppel, Karen Emmorey, Gregory Hickok, and Liina Pylkkänen. Towards a new neurobiology of language. *Journal of Neuroscience*, 32(41):14125–14131, 2012.

13. John T Hale, Luca Campanelli, Jixing Li, Shohini Bhattasali, Christophe Pallier, and Jonathan R Brennan. Neurocomputational models of language processing. *Annual Review of Linguistics*, 8:427–446, 2022.

14. Mostafa Abdou. Connecting neural response measurements & computational models of language: a non-comprehensive guide. *arXiv preprint* arXiv:2203.05300, 2022.

15. Tianyi Zhang and Tatsunori B. Hashimoto. On the inductive bias of masked language modeling: From statistical to syntactic dependencies. In *Proceedings of the 2021 Conference of the North American Chapter of the Association for Computational Linguistics: Human Language Technologies*, pages 5131–5146, Online, June 2021. Association for Computational Linguistics. https://doi.org/10.18653/v1/2021.naacl-main.404. https://aclanthology.org/2021.naacl-main.404.

16. Jennifer C. White and Ryan Cotterell. Examining the inductive bias of neural language models with artificial languages. In *Proceedings of the 59th Annual Meeting of the Association for*

Computational Linguistics and the 11th International Joint Conference on Natural Language Processing (Volume 1: Long Papers), pages 454–463, Online, August 2021. Association for Computational Linguistics. https://doi.org/10.18653/v1/2021.acl-long.38. https://aclanthology. org/2021.acl-long.38.

17. Bing Liu, Gokhan Tür, Dilek Hakkani-Tür, Pararth Shah, and Larry Heck. Dialogue learning with human teaching and feedback in end-to-end trainable task-oriented dialogue systems. In *Proceedings of the 2018 Conference of the North American Chapter of the Association for Computational Linguistics: Human Language Technologies, Volume 1 (Long Papers)*, pages 2060–2069, New Orleans, Louisiana, June 2018a. Association for Computational Linguistics. https://doi.org/10.18653/v1/N18-1187. https://aclanthology.org/N18-1187.

18. Ashwin Paranjape and Christopher Manning. Human-like informative conversations: Better acknowledgements using conditional mutual information. In *Proceedings of the 2021 Conference of the North American Chapter of the Association for Computational Linguistics: Human Language Technologies*, pages 768–781, Online, June 2021. Association for Computational Linguistics. https://doi.org/10.18653/v1/2021.naacl-main.61. https://aclanthology.org/2021.naacl-main.61.

19. Frank Keller. Cognitively plausible models of human language processing. In *Proceedings of the 48th Annual Meeting of the Association for Computational Linguistics: Short Papers*, pages 60–67, 2010.

20. Peter Lindes and John E Laird. Toward integrating cognitive linguistics and cognitive language processing. In *Proceedings of the 14th International Conference on Cognitive Modeling (ICCM)*, 2016.

21. Alexis Conneau, German Kruszewski, Guillaume Lample, Loïc Barrault, and Marco Baroni. What you can cram into a single $&!#* vector: Probing sentence embeddings for linguistic properties. In *Proceedings of the 56th Annual Meeting of the Association for Computational Linguistics (Volume 1: Long Papers)*, pages 2126–2136, Melbourne, Australia, July 2018. Association for Computational Linguistics. https://doi.org/10.18653/v1/P18-1198. https://aclanthology. org/P18-1198.

22. Yair Lakretz, Théo Desbordes, Dieuwke Hupkes, and Stanislas Dehaene. Can transformers process recursive nested constructions, like humans? In *Proceedings of the 29th International Conference on Computational Linguistics*, pages 3226–3232, Gyeongju, Republic of Korea, October 2022. International Committee on Computational Linguistics. https://aclanthology.org/2022. coling-1.285.

23. Wasi Ahmad, Zhisong Zhang, Xuezhe Ma, Eduard Hovy, Kai-Wei Chang, and Nanyun Peng. On difficulties of cross-lingual transfer with order differences: A case study on dependency parsing. In *Proceedings of the 2019 Conference of the North American Chapter of the Association for Computational Linguistics: Human Language Technologies, Volume 1 (Long and Short Papers)*, pages 2440–2452, Minneapolis, Minnesota, June 2019. Association for Computational Linguistics. https://doi.org/10.18653/v1/N19-1253. https://aclanthology.org/N19-1253.

24. Koustuv Sinha, Prasanna Parthasarathi, Joelle Pineau, and Adina Williams. UnNatural Language Inference. In *Proceedings of the 59th Annual Meeting of the Association for Computational Linguistics and the 11th International Joint Conference on Natural Language Processing (Volume 1: Long Papers)*, pages 7329–7346, Online, August 2021. Association for Computational Linguistics. https://doi.org/10.18653/v1/2021.acl-long.569. https://aclanthology.org/2021.acl-long.569.

25. Koustuv Sinha, Amirhossein Kazemnejad, Siva Reddy, Joelle Pineau, Dieuwke Hupkes, and Adina Williams. The curious case of absolute position embeddings. In *Findings of the Association for Computational Linguistics: EMNLP 2022*, pages 4449–4472, Abu Dhabi, United Arab

Emirates, December 2022. Association for Computational Linguistics. https://aclanthology.org/2022.findings-emnlp.326.

26. Pratik Joshi, Sebastin Santy, Amar Budhiraja, Kalika Bali, and Monojit Choudhury. The state and fate of linguistic diversity and inclusion in the NLP world. In *Proceedings of the 58th Annual Meeting of the Association for Computational Linguistics*, pages 6282–6293, Online, July 2020. Association for Computational Linguistics. https://doi.org/10.18653/v1/2020.acl-main.560. https://aclanthology.org/2020.acl-main.560.